Electrochemical Impedance Spectroscopy and Related Techniques

From Basics to Advanced Applications

Advanced Textbooks in Chemistry

Print ISSN: 2059-7673
Online ISSN: 2059-7681

The *Advanced Textbooks in Chemistry* series explores key topics in Chemistry that all Physical Sciences postgraduate students need to know to know to pass their MSc or PhD.

Written by senior academics as well lecturers recognised for their teaching skills, they offer in around 200 to 250 pages a theoretical overview of modern concepts.

Their lively style, focused scope and pedagogical material make them ideal learning tools at a very affordable price.

Most authors are based at prestigious universities: Imperial College London, Oxford, UCL, Ecole Polytechnique.

Published

Electrochemical Impedance Spectroscopy and Related Techniques:
From Basics to Advanced Applications
   by Laurence M Peter (University of Bath, UK)

Introduction to Heterogeneous Catalysis
(Second Edition)
   by Roel Prins (Institute for Chemical and Bioengineering, ETH Zurich, Switzerland), Anjie Wang (Dalian University of Technology, China), Xiang Li (Tianjin University of Science and Technology, China) and Foteini Sapountzi (Syngaschem BV, The Netherlands)

Introduction to Heterogeneous Catalysis
   by Roel Prins (Institute for Chemical and Bioengineering, ETH Zurich, Switzerland), Anjie Wang (Dalian University of Technology, China), and Xiang Li (Dalian University of Technology, China)

Advanced Textbooks in Chemistry

Electrochemical Impedance Spectroscopy and Related Techniques

From Basics to Advanced Applications

Laurence M Peter

University of Bath, UK

World Scientific

NEW JERSEY • LONDON • SINGAPORE • BEIJING • SHANGHAI • HONG KONG • TAIPEI • CHENNAI • TOKYO

Published by

World Scientific Publishing Europe Ltd.

57 Shelton Street, Covent Garden, London WC2H 9HE

Head office: 5 Toh Tuck Link, Singapore 596224

USA office: 27 Warren Street, Suite 401-402, Hackensack, NJ 07601

Library of Congress Cataloging-in-Publication Data

Names: Peter, Laurence (Laurence M.), author.
Title: Electrochemical impedance spectroscopy and related techniques : from basics to
    advanced applications / Laurence M. Peter, University of Bath, UK.
Description: New Jersey : World Scientific, [2024] | Series: Advanced textbooks in chemistry,
    2059-7673 | Includes bibliographical references and index.
Identifiers: LCCN 2023023566 | ISBN 9781800614505 (hardcover) |
    ISBN 9781800614512 (ebook for institutions) | ISBN 9781800614529 (ebook for individuals)
Subjects: LCSH: Impedance spectroscopy--Textbooks. | Impedance spectroscopy--
    Experiments. | Electrochemical analysis--Experiments.
Classification: LCC QD116.I57 P48 2024 | DDC 535.8/4--dc23/eng/20230928
LC record available at https://lccn.loc.gov/2023023566

British Library Cataloguing-in-Publication Data

A catalogue record for this book is available from the British Library.

For any available supplementary material, please visit
https://www.worldscientific.com/worldscibooks/10.1142/Q0428#t=suppl

Desk Editors: Nambirajan Karuppiah/Ana Ovey/Shi Ying Koe

Typeset by Stallion Press
Email: enquiries@stallionpress.com

For Renate

Preface

The idea for this book grew out of a long-running annual course on electrochemical impedance spectroscopy (EIS) held at the University of Bath. Each year, Bath welcomes course participants with a range of different backgrounds to a four-day intensive EIS course that combines lectures with practical hands-on experience using modern equipment. Achieving an in-depth understanding of impedance spectroscopy in just four days is highly demanding, particularly for those who are new to the topic. This book was initially conceived as an opportunity to address this problem by taking a more relaxed in-depth look at the basic principles of EIS before moving on to illustrate the applications of EIS using experimental examples taken from the extended laboratory classes that form an essential component of the Bath course. To help readers come to grips with the basic concepts of impedance, the text includes detailed *Theory Notes* that provide a framework for understanding EIS that goes beyond equivalent circuits and fitting software. However, as work on the book progressed, it became clear that its initial scope was too limited. Consequently, additional chapters have been included to extend the concepts of frequency-response analysis to other small-amplitude frequency-resolved methods, including intensity-modulated photocurrent/photovoltage spectroscopy (IMPS/IMVS) and potential/light-modulated optical absorbance/reflectance (PMAS/LMAS), techniques that are widely used in solar cell and solar water-splitting research. In its final form, the book takes the reader from basic beginnings to more recent applications that illustrate the universal utility of frequency-resolved techniques across a range of research areas. Some applied topics such as batteries and corrosion have been omitted since there are already books and review papers that cover these in detail.

About the Author

Laurence M Peter is an Emeritus Professor at the University of Bath in the UK. Originally trained as an electrochemist, his interests have diversified over the years to include novel photovoltaics and green hydrogen generation by light-driven water splitting. He is best known for his development of novel frequency-resolved optoelectronic techniques to study semiconductor electrodes and solar cells. His work in this area has been recognized by the award of the Pergamon Prize of the International Society of Electrochemistry and the Electrochemistry Prize of the Royal Society of Chemistry. Around 30 years ago, he founded the annual Bath course in Electrochemical Impedance Spectroscopy. Combining lectures with extended lab sessions, this highly popular hands-on course continues every year to introduce scientists from industry and academia to experimental techniques and background theory in EIS. The author's research as well as teaching experience gained from running the EIS course provided the motivation for this book.

Acknowledgements

I am immensely grateful to my colleagues from the University of Bath (Petra Cameron, Sarah Dale, Toby Jenkins, Frank Marken), the University of Bristol (David Fermin), Imperial College (Jason Riley) and Solartron Analytical (John Harper, Andrew Savage), for their input to the Bath EIS course over many years. Their efforts and encouragement provided the motivation for this book, but any errors, omissions or inconsistencies in the text are entirely my responsibility. I would like to thank David Fermin for coming up with the original idea of writing a book based on the Bath EIS course and Frank Marken for providing experimental data and outline text, particularly for Chapter 6. Petra Cameron and Adam Pockett kindly let me use some of their EIS and IMVS data on planar perovskite solar cells, and Charles Cummings, Frank Marken, Upul Wijayantha and Asif Tahir also allowed me to analyze data that we obtained in our joint study of light-driven water splitting on haematite electrodes. I am also grateful to Scribner Associates (Derek Johnson) for kind permission to generate figures using ZView® impedance fitting software and to Michael Schönleber (Batemo GmbH) for permission to use the Lin–KK program in the chapter on Kramers–Kronig analysis of impedance data. Gamry Instruments kindly permitted me to adapt the picture of their excellent rotating disc setup for the figure in Chapter 6. John Harper (Solartron Analytical) kindly agreed to let me use a figure from Solartron documentation. It was also a pleasure to work with him on the development of the Solartron ModulabXM system, which was used to generate many of the experimental results in the book. My thanks also go to Evgueni Ponomarev, who kindly checked my derivations in the appendix to Chapter 10 and to Laurent Chaminade (World Scientific Publishing), who kept faith in the project when it seemed that the impact of COVID-19 could make it impossible. I apologize if I have forgotten anyone else who has helped make this book possible. Finally, my special thanks to my wife, Renate, to whom this book is dedicated.

Contents

Theory Notes

List of Symbols

A	electrode area
α	absorption coefficient or transfer coefficient
C	capacitance or concentration
C_{dl}	differential double-layer capacitance
C_H	Helmholtz capacitance
C_{sc}	differential space charge capacitance
$\hat{C}$	complex capacitance
d	layer thickness
δ	diffusion layer thickness
D	diffusion coefficient
E	electrode potential
E_{fb}	flat band potential
ε	relative permittivity
ε_0	permittivity of free space
f	frequency
F	Faraday constant
i	current
i_0	exchange current
I_0	incident photon flux density
j	$\sqrt{-1}$ or current density
δj	small amplitude modulated current density
j_G	hole current density predicted by the Gärtner equation
j_{rec}	recombination current density

k_B	Boltzmann constant
k_{trans}	first-order charge transfer rate constant
k_{rec}	pseudo-first-order recombination rate constant
L	inductance
L_p	hole diffusion length
L	Laplace transformation
N_a	acceptor doping concentration
N_d	donor doping concentration
n_{surf}	concentration of electrons at the surface
ω	radial frequency
ω_{rot}	rotation rate of RDE in radians s^{-1}
p	surface hole concentration
q	elementary charge
Q	charge density
δQ	small amplitude modulated charge density
R	resistance or the gas constant
R_{ct}	charge transfer resistance in distributed element
R_F	Faradaic resistance
T	absolute temperature
τ	charge carrier lifetime
τ_{diff}	diffusion time constant
V	voltage
δV	small amplitude modulated voltage
W_{sc}	thickness of space charge layer
Y	admittance
Z	impedance
Z_{sc}	space charge impedance
Z_H	Helmholtz impedance

Chapter 1

Getting Started

1.1 Introduction

This book traces a path from the basic concepts of impedance spectroscopy to more advanced applications of frequency-resolved techniques. Trying to travel the whole path in one go is not something that most people will want to do. Instead, it is best to break the journey up into manageable sections. If you are new to impedance spectroscopy, some later sections of the book may look too difficult at first. If this is the case, you can focus on the easier parts until you have summoned up enough expertise and courage to tackle later chapters. If, on the other hand, you are already familiar with the basic concepts of impedance spectroscopy, you should be able to progress more rapidly through the first couple of chapters.

Every journey requires some preparation, and this chapter is intended to give you enough background to understand the basic concepts that we will use when discussing impedance or *electrochemical impedance spectroscopy* (EIS). We are going to need some theory, but I have tried to introduce the necessary concepts as gently as possible, so don't panic if you have an aversion to mathematics or if you have forgotten what you learned at school, college or university. Things should become clearer as you progress. Some of the ideas may seem rather abstract at first, but if you persevere, you should get a real grasp of what is involved in the concept of impedance and be able to deal confidently with the treatment and analysis of impedance data. At the end of the chapter, you will find some exercises that will allow you to check if you are ready to progress to Chapter 2. It is important to spend some time becoming familiar with the basic ideas since they underpin all the diverse material that is covered in the following chapters.

1.2 Inputs, Outputs, Transfer Functions and Linear Circuit Elements

Perhaps you remember being given a mysteriously wrapped present as a child and shaking it to guess what was inside. If you do remember, you probably did not realize that you were carrying out a scientific experiment involving *perturbation* of a system and observation of the *response*. For physical systems, we can think of a large range of possible kinds of perturbations and responses. A few are summarized in Table 1.1. In all these cases, we apply some perturbation or stimulus and get some response.

Table 1.1. Some examples of perturbations (inputs) and responses (outputs).

System or device	Perturbation	Response
Electrical/electrochemical	Voltage	Current
Electrical/electrochemical	Current	Voltage
Photodiode	Light	Current
Thermocouple	Temperature	Voltage
Pressure transducer	Pressure	Voltage
Microphone	Sound	Voltage
Ion channel	Nerve stimulus	Voltage

The outputs of these systems are all related to the inputs by a *transfer function*. The inputs we are interested in are all small amplitude sinusoidal signals, and the outputs are also sinusoidal. When we apply a sinusoidal signal to a system, we generally see a *transient response* before the response settles down after a few cycles to a steady *periodic response*. In EIS and related techniques, it is the periodic response that we focus on. For the moment we will use the term 'transfer function' rather loosely to mean any relationship that relates steady state periodic input and output signals. In electronics textbooks, you will find that transfer functions are defined in terms of something called the *complex frequency variable* that we will meet later in this chapter.

The first kind of transfer function we will meet is the *impedance, Z*, the main subject of much of this book. Suppose we have a two-terminal 'black box' (Figure 1.1) containing a network consisting of *linear circuit elements* (resistors, capacitors and inductors).

For the moment, we define the impedance Z in terms of the voltage V across the box and the current i through it.

$$Z = \frac{V}{i} \tag{1.1}$$

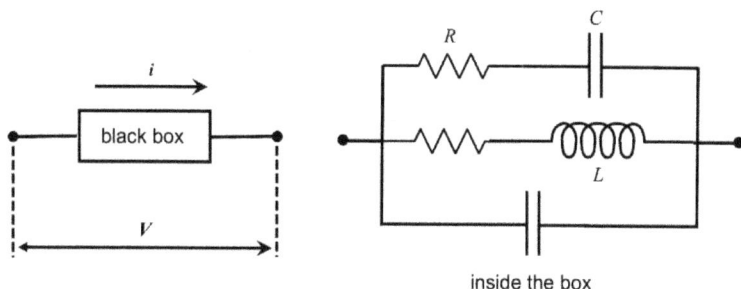

inside the box

Figure 1.1. Defining impedance and admittance in terms of the voltage, V, applied across the terminals of a 'black box' and the current, i, flowing between the terminals. The black box contains a network of *linear circuit elements*, which can be resistors (R), capacitors (C) and inductors (L). In impedance spectroscopy, both V and i are sine waves of the same frequency, but generally with different *phases*, i.e., the peaks are shifted in time relative to one another. The circuit example on the right illustrates the three types of linear circuit elements.

Later we shall come across the square root of -1, which is also given the symbol i in mathematics, so to avoid possible confusion, we follow the example of electrical engineers and use the symbol j for $\sqrt{-1}$ throughout this book. The units of Z are ohms (Ω). In impedance spectroscopy, both V and i are not constant but vary periodically with time. The general form of V that we use in EIS is a small amplitude *sine wave*. The current i has the same frequency as the voltage V, but it may be shifted on the time axis relative to the voltage. This means that to specify V and i, we need to provide information not only about the *amplitude* of the sinusoidal oscillations but also about something called the *phase*, which describes how the waveform is shifted on the time axis relative to another wave of the same frequency. Equation (1.1) is therefore not as straightforward as it first appears. We need to have some way of accounting for the fact that V and i need to be specified in terms of their magnitude and their phase. This is what we will explore in this chapter.

The admittance, Y, is defined as the inverse of the impedance, i.e.,

$$Y = \frac{1}{Z} = \frac{i}{V} \tag{1.2}$$

Y, therefore, has units of Ω^{-1} or, in SI units, siemens (capital S). In some older books, you may find mho (ohm back to front) instead of S. Again, equation (1.2) appears to be simple, but remember that we are dealing with voltage and current signals that are likely to be shifted on the time axis relative to each other.

We can also have 'black boxes' for which the input and outputs are both voltages, or are both currents. Such systems can be amplifiers (they make the voltage or current signal bigger) or *attenuators* (they make the signals smaller). In such cases, we can have dimensionless transfer functions (i.e., the units of the nominator and denominator cancel out) such as

$$H_V = \frac{V_{out}}{V_{in}} \quad H_i = \frac{i_{out}}{i_{in}} \tag{1.3}$$

where H_V is a voltage transfer function, and H_i is a current transfer function. Again, the relative phases of the voltages or currents are important and need to be accounted for when defining the transfer functions.

Let us take another example from Table 1.1. For a photodiode, the input signal is a photon flux I_{photon} (photons s^{-1}) and the output is an electron flux (electrons s^{-1}) or current i_{diode} (amps = coulombs s^{-1} : A $\equiv$ Cs^{-1}). In this case, we can define a dimensionless photodiode transfer function by converting the current to an electron flux by dividing i_{diode} by the elementary charge, q, to obtain

$$H_{pd} = \frac{\text{Electron flux out}}{\text{Photon flux in}} = \frac{i_{diode}}{qI_{photon}} \tag{1.4}$$

where q is the elementary charge (1.602×10^{-19} C).

EIS generally deals with responses that are related linearly to the perturbation. An example of this is the familiar relationship between voltage and current for a resistor – Ohm's law:

$$V = iR \tag{1.5}$$

Ohm's law tells us that when we vary the voltage, V (volts), the current, i (amps), responds linearly, with the coefficient of proportionality being the resistance, R, which is measured in ohms (Ω). We call R a *linear circuit element*. You can see that our definition of impedance in equation (1.1) looks like Ohm's law, except that R is replaced by Z.

There are two other types of linear circuit elements shown in Figure 1.1. The first is the capacitor, C. This is an electronic component that can store charge on two conducting plates separated by an insulating dielectric material. The charge, Q, (coulombs) stored in a capacitor is a linear function of the applied voltage, V, which we can express as:

$$Q = CV \tag{1.6}$$

The coefficient of proportionality is the capacitance, C, which is measured in units of farads ($F \equiv CV^{-1}$). The capacitance of a parallel plate capacitor is determined by A, the area of the plates, d, their separation and ε, the relative permittivity (dielectric constant) of the material between them, as follows:

$$C = A\frac{\varepsilon\varepsilon_0}{d} \tag{1.7}$$

Here, ε_0 is the *permittivity of free space* (8.854×10^{-14} F cm^{-1}). In electrochemistry, C_{dl}, the differential *double-layer capacitance* is an important element in fitting EIS data. In this case, one of the charged plates is the electrode, and the other 'plate' consists of ions in solution that form the *electrical double layer*.

If the voltage V changes periodically with time, we can take the time derivative d/dt of both sides of equation (1.6) to discover that the current is proportional to the rate of change of voltage (remember that charge per second = current: Cs^{-1} ≡ A), as follows:

$$\frac{dQ}{dt} = i = C\frac{dV}{dt} \tag{1.8}$$

i.e., the current through a capacitor is proportional to the rate of change of voltage with time. If the voltage does not vary with time, $dV/dt = 0$, and the current is therefore also zero. This means that a capacitor blocks dc current but allows ac current to pass. If the time-dependent voltage is a sine wave, its derivative dV/dt will be a cosine wave. This means that the voltage across the capacitor and the current through it are not in phase, but instead are phase shifted by 90° with respect to each other.

The last linear circuit element is the inductor, L. It typically consists of a coil of low-resistance wire wound around a magnetic (e.g., iron or ferrite) core. When a time-varying current flows through the coil, it produces a time-varying magnetic field that generates an electromotive force or voltage that opposes the change in current. The voltage drop across the inductor depends on the rate of change of current and is given by the linear relationship

$$V = L\frac{di}{dt} \tag{1.9}$$

The coefficient of proportionality, L, is the inductance, measured in Henrys (H ≡ Ωs ≡ VA^{-1}s). You can see that the voltage drop across an inductor should be zero in the case of a dc current since $di/dt = 0$. This means that the dc impedance of an ideal inductor is also zero. However, real inductors have a finite resistance due to the wire, so that there is a finite voltage drop across them, even for a dc current. The $\frac{di}{dt}$ term tells us that the current and voltage will again be shifted relative to each other by 90°. More of this later. In practice, inductors are used to block ac signals, for example in dc power supplies, which is why they are also called chokes.

1.3 Nonlinear Systems

We now take a quick look at some electrochemistry to illustrate how EIS can handle nonlinear systems. In the case of electrochemical systems, the current generally does not vary linearly with voltage. In fact, we often find that the current increases *exponentially* with voltage as shown in Figure 1.2. Electrochemists express this nonlinear relationship as the *Tafel equation* (named after Julius Tafel, a Swiss chemist and early electrochemist), as follows:

$$i = i_0 e^{\frac{\alpha n F(E - E_{eq})}{RT}} \tag{1.10}$$

Here, i_0 is the exchange current, n is the number of electrons transferred in the reaction, α is the transfer coefficient, F is the Faraday constant, $(E - E_{eq})$ is the voltage applied relative to the equilibrium potential, E_{eq}, R is the gas constant, and T is the absolute temperature.

In general, we can get around the problem of the nonlinear input–output relationships by making the perturbation very small, as shown in Figure 1.2. Remaining with the Tafel equation, let us suppose that we apply a dc voltage V_{dc} (relative to the equilibrium potential) to an electrode in our electrochemical system. We obtain some dc current i_{dc}. Now we 'tickle' the system with a small additional voltage δV. This increases the current by a small amount δi. Figure 1.2 shows that we can consider the small section of the curve involved in the increase as linear. In other words, the input δV and output δi are linearly related to a good approximation. In EIS, δV and δi are ac signals.

Let us see how the linearization illustrated in Figure 1.2 can be expressed mathematically. This is the first of the *Theory Notes* that you will find throughout this book. You can skim through them on a first reading if you wish, but I recommend that you return to them as often as necessary to make sure you understand them, because they provide the essential framework behind EIS.

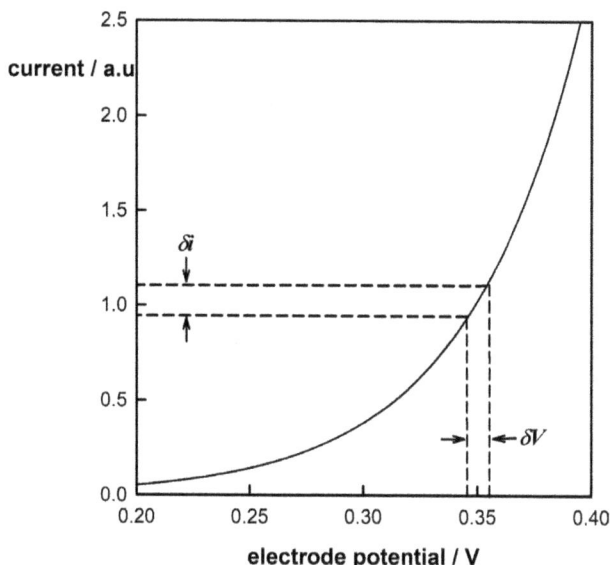

Figure 1.2. A plot of the Tafel equation showing the exponential increase in current with voltage (current scale in arbitrary units). We can study the system at any point on the curve (i.e., at any applied dc voltage V_{dc}) by adding a small voltage perturbation δV and measuring the change in current δi. Provided δV is small enough (a few mV), the input and output are linearly related to a good approximation.

Theory Note 1.1. Linearizing the Tafel Equation

To simplify things, let us replace the exponent in the Tafel equation (equation (1.10)) by fV where $V = (E - E_{eq})$ is the voltage and $f = \alpha n F/RT$. Our simplified Tafel equation is now

$$i_{dc} = i_0 e^{fV} \tag{1.11}$$

Let us now increase V by a small amount δV. Using the property of exponentials that $e^{a+b} = e^a e^b$, the current increases to a slightly higher value given by

$$i_{dc} + \delta i = i_0 e^{f(V+\delta V)} = i_0 \left[e^{fV} e^{f\delta V} \right] \tag{1.12}$$

Now for the linearization. An exponential function e^x can be expressed as a power series.

$$e^x = 1 + x + \frac{x^2}{2} + \frac{x^3}{6} + \frac{x^4}{24} + \cdots \tag{1.13}$$

(Continued)

Theory Note 1.1. (*Continued*)

If the exponent x is much less than 1, the series converges very rapidly, so to a good approximation we can replace e^x by $1 + x$ because higher powers of x become vanishingly small. If $f\delta V$ is small enough (in practice, we typically use $\delta V \approx 1 - 5\,\text{mV}$), we can replace $e^{f\delta V}$ in equation (1.12) by $(1 + f\delta V)$. The current at $V + \delta V$ is therefore

$$i_{dc} + \delta i \cong i_0 e^{fV} e^{f\delta V} = i_0 e^{fV} (1 + f\delta V) = i_0 e^{fV} + i_0 e^{fV} f\delta V \quad (1.14)$$

But the first term, $i_0 e^{fV}$, is equal to the current i_{dc} at V, so that in the case where δV is an ac signal

$$i_{dc} + \delta i = i_{dc} + i_{ac} \quad (1.15)$$

and therefore

$$\delta i = i_{dc} f \delta V \quad (1.16)$$

You can see that we have managed to obtain a linear relationship between the small change in current δi and the small change in voltage δV. In fact, we can write the result in the same form as Ohm's law $V = iR$.

$$\delta V = \frac{1}{i_{dc}f}\delta i = \frac{1}{\frac{i_{dc}\alpha nF}{RT}}\delta i = \frac{RT}{i_{dc}\alpha nF} \quad (1.17a)$$

$$\delta i = \delta i R_F \quad (1.17b)$$

where the *Faradaic resistance*, R_F, is given by

$$R_F = \frac{RT}{i_{dc}\alpha nF} \quad (1.18)$$

What about other nonlinear responses? In general, we can use a series expansion to represent the response and then take the linear terms for small changes just like we did for the exponential series, which is a special case of the more general *Taylor series* that you can find in Maths textbooks.

1.4 Sine and Cosine Waves

Impedance spectroscopy is really all about sine waves, even though we rarely see them when making impedance measurements with a modern *frequency-response analyzer* (FRA). We are all familiar with the idea that the mains ac voltage is a sine wave, but why is this? In the USA in the late 1880s, three larger-than-life characters – Thomas Edison on one side, George

Westinghouse and Nicola Tesla on the other – were engaged in a fierce battle over whether direct or alternating current (dc or ac) electricity held the key to the future. In the end, Westinghouse and Tesla won the 'war of the currents', and today the world is mainly powered by ac, which is basically generated by rotating a loop or coil of wire in a magnetic field between the poles of a magnet in a device called an alternator. The *frequency* of the resulting electromotive force (emf) is determined by the rotation rate of the generator, and is either 50 Hz (Hertz) or, in the USA, 60 Hz (the Hertz unit is named after the German polymath and physicist Heinrich Rudolf Hertz, who was adept in Arabic and Sanskrit as well as science). The waveform of the emf or voltage is determined by the orientation of the wire loop relative to the magnetic field lines that run from one pole of the magnet to the other. When the loop is parallel to the lines, the voltage is zero, and when it is perpendicular, the voltage is a maximum (either positive or negative). So as the coil rotates, it generates the familiar voltage waveform shown in Figure 1.3.

Nowadays, virtually all electronic devices work on dc, so our houses and offices are full of devices to convert ac to dc. Going the other way, devices like solar panels that generate dc power need to be connected to the electricity grid via devices called power inverters that convert dc to ac and are synchronized with the ac mains voltage.

So much for the history. Since we are talking rather loosely about 'sine waves', we need to take a closer look at what we mean by a sine wave. This involves some trigonometry that you may have forgotten from your school days. Time for another theory note.

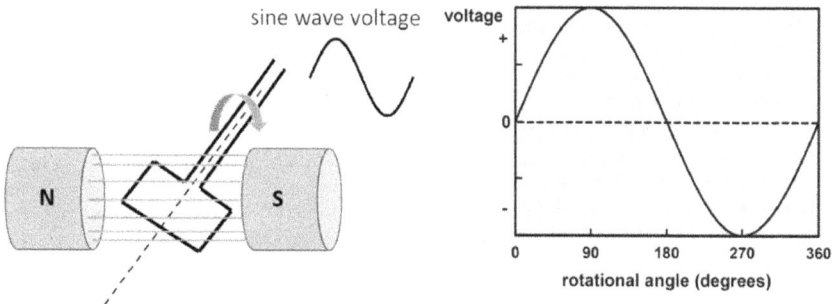

Figure 1.3. Generation of a sine wave voltage in an alternator. The voltage depends on the *sine* of the angle between the coil and the magnetic field lines.

Theory Note 1.2. Sines, Cosines and Tangents

Let us remind ourselves of what a *sine* is. The sine of an angle is defined using a right-angled triangle. It is given by $\sin \theta = $ opposite/ hypotenuse $= y/z$ as shown in Figure 1.4 (the hypotenuse is the longest side). The other definitions that we shall need later are for *cosine* and *tangent*: $\cos \theta = x/z$ and $\tan \theta = y/x$.

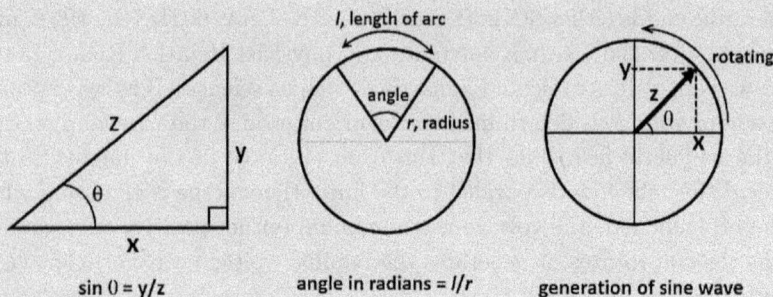

$$\sin \theta = y/z \qquad \text{angle in radians} = l/r \qquad \text{generation of sine wave}$$

Figure 1.4. Some definitions. Left: sine of an angle. Centre: angles in radians. Right: generation of sine wave by rotating the hypotenuse z (shown as an arrow) anticlockwise.

You will be familiar with measuring angles in degrees. We can also measure them in *radians* using a circle to define our angle. The angle in radians is the length of the arc divided by the radius. Remembering that the circumference of a circle is given as π times the diameter or 2π times the radius, we can see that a rotation of $360°$ corresponds in radians to the circumference of the circle divided by its radius, i.e., to 2π radians.

To generate a sine wave, imagine we fix the triangle in Figure 1.4 at its left-hand end and *rotate the hypotenuse* (z, shown by an arrow on the right-hand side of Figure 1.4) anticlockwise while allowing the lengths y and x to change to maintain the right-angled triangle shape. Let us start from an angle of $\theta = 0°$, which means the hypotenuse is horizontal (i.e., lying along the x axis), and then sweep the hypotenuse *anticlockwise* around a full circle so that θ increases from $0°$ to $360°$ (rotation 0 to 2π radians). The sine of the angle θ then follows the 'sine wave' shape shown in Figure 1.3.

If our arrow rotates f times a second, the frequency is f hertz. When discussing impedance, we often use the *radial frequency*, ω, which is the number of radians per second that the arrow passes through. Since one rotation of 360° corresponds to 2π radians, we see that

$$\omega = 2\pi f \tag{1.19}$$

and the variation of θ (in radians) with time, $\theta(t)$, is therefore

$$\theta(t) = 2\pi f t = \omega t \tag{1.20}$$

Looking at the right-hand side of Figure 1.4, you can see that when $\theta = 0$, the arrow points along the x axis to the right, i.e., $y = 0$, so $\sin\theta = y/z = 0$. As the arrow rotates anticlockwise, its length z remains constant, but y increases until it equals z when the arrow points up the y axis and then decreases again to zero when the arrow points along the x axis towards the left. y then changes sign and reaches $-z$ when $\theta = 270°(3\pi/2$ radians) before the arrow completes the rotation, and y returns to zero. Remembering that $\sin\theta = y/z$, you should be able to work out for yourself how the rotating arrow produces the familiar sine wave shown in Figure 1.3.

So far, we have just talked about sine waves in general terms. However, in EIS we are interested in voltages and currents rather than just trigonometric functions. For the following discussion, we are going to use cosines rather than sines. The cosine of θ is given by $\cos\theta = $ adjacent/hypotenuse $= x/z$. Look at the length of the adjacent side, x, of the triangle made by the rotating arrow in Figure 1.4. It is equal to the vertical projection from the tip of the arrow onto the x axis. Now suppose that the length of the arrow represents a voltage V_o. As the line rotates anticlockwise round the circle from 0 to 2π radians, the projection of the tip of the arrow onto the x axis starts from V_o, passes through zero when $\theta = \pi/2(90°)$, then reaches $-V_o$ at $\theta = \pi(180°)$ before passing through zero again when $\theta = 3\pi/2(270°)$ and returning to V_o after one full cycle. This projection on the x axis corresponds to our time-varying ac voltage, $V(t)$. Since $\cos\theta = \frac{x}{V_0} = V_t/V_0$, you can see that our ac voltage is a cosine wave.

$$V(t) = V_0 \cos[\theta(t)] = V_0 \cos(\omega t) \tag{1.21}$$

From now on, we will use equation (1.21) to represent the time dependence of our ac voltage. So, our sine wave has become a cosine wave. A quick reminder of the difference (Figure 1.5).

angle in radians

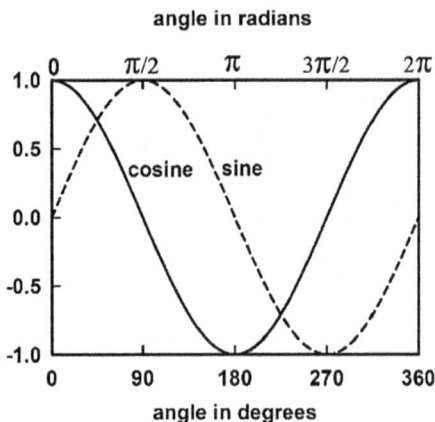

angle in degrees

Figure 1.5. Comparison of sine and cosine functions. Note that the sine wave plot is displaced 90° ($\pi/2$ radians) to the right relative to the cosine plot.

1.5 Currents, Voltages, Phases and Roundabouts

Let us suppose we have a mystery black box with an input terminal and an output terminal. The input is a time-dependent cosine voltage applied between the terminals and the output is the current flowing through the box. Suppose, for example, that we find that the current signal has the same sinusoidal shape as the voltage signal but is shifted along the time axis as shown in Figure 1.6.

The output current signal contains two kinds of information. First, how big it is (the *amplitude*). Second, how it is shifted along the time axis relative to the input voltage signal. Now remember that the time axis is

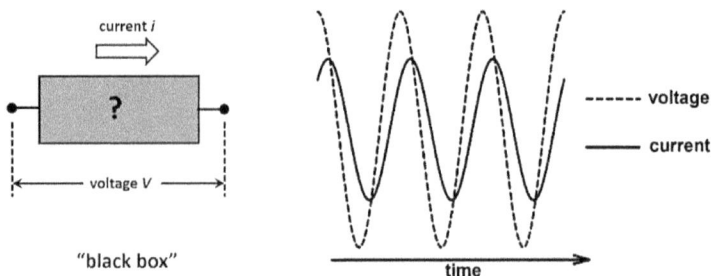

Figure 1.6. Input voltage and output current from a black box measured at a particular frequency. Note that the peaks in the two signals do not coincide. The current signal is *phase shifted* relative to the voltage signal.

also an angle axis because one full cycle of the signal corresponds to 360°
or 2π radians. This means that we can express the shift of the current trace
along the time axis in terms of a shift in angle – this is called the *phase
shift*. You can see that the current trace in Figure 1.6 is shifted to the right
of the voltage by some fraction of the time between successive maxima in
the dashed line plot. Since the successive maxima are separated by 360°
or 2π radians, we can work out the phase angle difference between voltage
and current from the time difference between the current and voltage peaks.
Figure 1.7 shows how this can be done using an oscilloscope trace. In the
(arbitrary) example shown in the figure, the phase shift is +45°($\pi/4$).

Now look back at Figure 1.4. Here we used the idea of an arrow rotating
at a fixed rate to generate an ac signal. We can use this idea to represent the
cosine (voltage) wave in Figure 1.7. But what about the current signal? It
has (i) the same frequency, (ii) an amplitude measured in amperes rather
than volts and (iii) a different phase angle. By this time, you may have
guessed that we can represent both the voltage and the current on the
same diagram with rotating arrows. The length of the arrows will show the
amplitudes (in the appropriate units of V and A), and the phase shift will
be the angle between the two lines as shown in Figure 1.8. Since the arrows
have both *magnitude* and *direction*, we refer to them as *vectors*. Rotating
vectors like those in Figure 1.8 are called *phasors*.

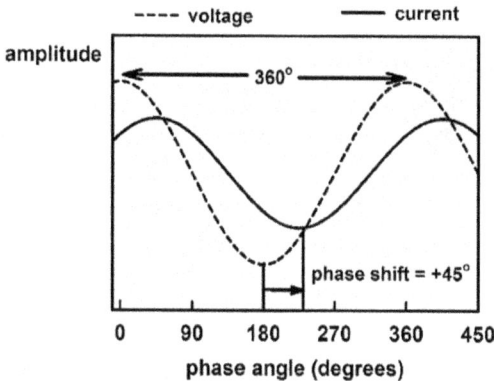

Figure 1.7. Working out the phase shift between current and voltage from an
oscilloscope trace. The phase angle shown on the x axis corresponds to the voltage
signal, which is a cosine function – see equation (1.21). Note that the positive peaks in
the voltage trace are 360° (2π radians) apart, allowing conversion of the x axis from
units of time to units of degrees or radians. The *phase shift* (in this case 45° or $\pi/4$) is
the difference between the minima (or maxima) in the current and the voltage signals.

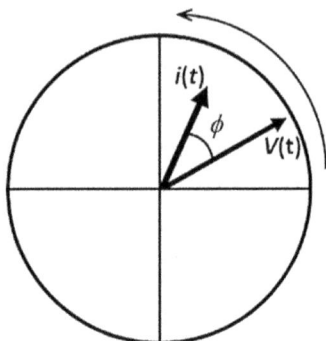

Figure 1.8. Representing voltage and current signals using rotating vectors or *phasors* (the arrows) rotating at a radial frequency $\omega = 2\pi f$. The amplitude of the signals is shown by the length of the arrows and the phase shift between them by ϕ. Imagine the hour and minute hands of a malfunctioning clock rotating anticlockwise at the same speed with a fixed angle between them.

At the beginning of this chapter, we highlighted the fact that when defining impedance or admittance, we needed to take both the magnitude and the phase of the current and voltage into account. The phasor representation shown in Figure 1.8 does exactly this. We are now getting to some very important concepts that involve some rather neat Maths. Time for another theory note dealing with *complex numbers*.

Theory Note 1.3. Entering the Imaginary j Realm

Now we are going to look at some smart maths dating back nearly 300 years to the great Swiss polymath *Leonhard Euler*. Widely considered one of the greatest mathematicians of all time, Euler's 866 publications will fill 81 volumes of the *Opera Omnia Leonhard Euler* when it is finished. Most of the notation we use every day in mathematics originated from Euler. The smart maths is related to the idea of the *square root of* -1.

The square root of -1? How can negative numbers have square roots? In fact, the idea is purely imaginary but incredibly useful. So useful, in fact, that the square root of -1 is given the symbol i. In electrical engineering and EIS, j is used instead of i.

$$j = \sqrt{-1} \tag{1.22}$$

Theory Note 1.3. (*Continued*)

You may already have come across complex numbers. These are written in the form:

$$z = x + jy \qquad (1.23)$$

The complex number z is made up of a real part, x, and an imaginary part, y.

We can represent any complex number using a graph. The x axis corresponds to the real part of the number, and the y axis to the imaginary part. These axes define the *complex plane*. A complex number is just a point (x, y) on this complex plane defined by its real (x) and imaginary (y) parts as shown on the left-hand side of Figure 1.9. We often draw an arrow from the origin $(0, 0)$ to the point (x, y).

So, what about Euler? He proved the following equation for any angle φ (in radians).

$$e^{j\varphi} = \cos\varphi + j\sin\varphi \qquad (1.24a)$$

and

$$e^{-j\varphi} = \cos\varphi - j\sin\varphi \qquad (1.24b)$$

This is *Euler's equation*.

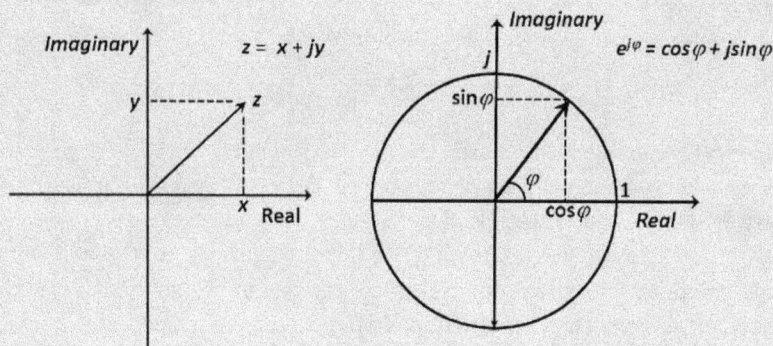

Figure 1.9. Left: Using the *complex plane* to represent a complex number $z = x + jy$ as a point (at the tip of the arrow). The real component of z is x, the projection on the real axis. The imaginary component of z is y, the projection on the imaginary axis. Right: Euler's formula shown in the complex plane using a circle with a radius of 1. The x-axis intercept corresponds to the real part and the y-axis intercept, to the imaginary part of the right-hand side of equation (1.24).

(*Continued*)

Theory Note 1.3. (*Continued*)

The right-hand side of Figure 1.9 shows how we can plot Euler's equation using a circle of radius 1 in the complex plane. Note that the y axis is the imaginary axis and that the circle crosses this axis at j because the circle radius is 1.

So, what happens if the arrow is a *rotating phasor*? We know that this generates a cosine wave, and there it is in the real part of Euler's equation (the sine wave is in the imaginary part). So, we can go back to our equation for the cosine voltage – equation (1.21) – and using Euler's equation, replace the cosine bit by the real part (Re) of a *complex exponential*

$$V(t) = V_0 \cos \omega t = V_0 \mathrm{Re}\left(e^{j\omega t}\right) \qquad (1.25)$$

But why would we want to do this? The answer can be found by looking at Figure 1.8, where the current signal happens to be shifted by $\phi = 45°$ relative to the voltage signal. Any phase shift ϕ in the current relative to the voltage can be represented by

$$i(t) = i_0 \cos(\omega t + \phi) \qquad (1.26)$$

How do we represent this current signal using Euler's equation? Easy – it is just like equation (1.25), but with V_0 replaced by i_0 and $j\omega t$ replaced by $j(\omega t + \phi)$.

Let us see what happens when we do this.

$$i(t) = i_0 \cos(\omega t + \phi) = i_0 \mathrm{Re}\left[e^{j(\omega t + \phi)}\right] = i_0 \mathrm{Re}\left[e^{j\phi}\right] \times \mathrm{Re}\left[e^{j\omega t}\right] \quad (1.27)$$

This was the smart bit. We used the property of exponentials $e^{ab} = e^a \times e^b$ to *separate out the rotating phasor part* – the $e^{j\omega t}$– from the phase shift part, ϕ. Danke schön, Herr Euler.

It may help to visualize what we have just done in the following way. Suppose you were looking at a playground roundabout rotating anticlockwise at a fixed frequency. Two bars on the roundabout appear to be at a fixed angle, but you are not sure because both are rotating, making you dizzy. Taking your life in your hands, you jump onto the roundabout. Now you can see the angle clearly and since you are also rotating at the same speed, you can forget about the fact that both bars are rotating. We got rid of the $e^{j\omega t}$ part. The $j\omega$ terms form the *complex frequency variable* that was mentioned in the introduction when defining what we mean by a transfer function. The transfer functions that we will encounter are functions of the complex frequency variable $j\omega$ (Figure 1.10).

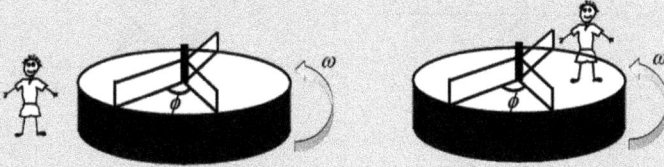

Figure 1.10. Freezing out the rotation $e^{j\omega t}$ by jumping on the roundabout.

Now that we have both current and voltage in complex exponential form, we are finally ready to properly define the impedance of our black box. When we defined the impedance, Z, in equation (1.1), we pointed out that we needed to think about magnitude and phase. We now know that both i and V are varying cosinusoidally with time with a phase difference ϕ. Using what we have discovered about complex exponentials, we can write the ac impedance Z as

$$Z = \frac{V(t)}{i(t)} = \frac{V_0 \cos(\omega t)}{i_0 \cos(\omega t + \phi)} = \frac{V_0 \mathrm{Re}\left[e^{j\omega t}\right]}{i_0 e^{j\phi}\mathrm{Re}\left[e^{j\omega t}\right]} = \frac{V_0}{i_0}e^{-j\phi} \qquad (1.28)$$

What has happened here? *The 'rotation' part –* $\mathrm{Re}\left[e^{j\omega t}\right]$ *– cancels out because the voltage and current phasors are both rotating at the same frequency,* leaving just two terms. The first term is the ratio of the *amplitudes* of the voltage and current signals, V_0/i_0, which defines the magnitude of the impedance and is written using vertical line brackets as $|Z|$. The second term is the complex exponential $e^{-j\phi}$. But we know from Herr Euler that we can replace the complex exponential $e^{-j\phi}$ by a real cosine term and an imaginary sine term. So, we find that the impedance is given by

$$Z = \frac{V_0}{i_0}e^{-j\phi} = |Z|e^{-j\phi} = |Z|[\cos(\phi) - j\sin(\phi)] \qquad (1.29)$$

Note that our impedance has a real part $|Z|\cos\phi$ and an imaginary part $-|Z|\sin\phi$. Let us look at this using a graph in the complex plane – Figure 1.11. At some point. electrical engineers, suffering from stiff necks from looking at upside-down complex plane impedance plots, adopted the upside-down 'Australian' convention with the $-j$ axis pointing up, and this is how impedance measurements are normally displayed.

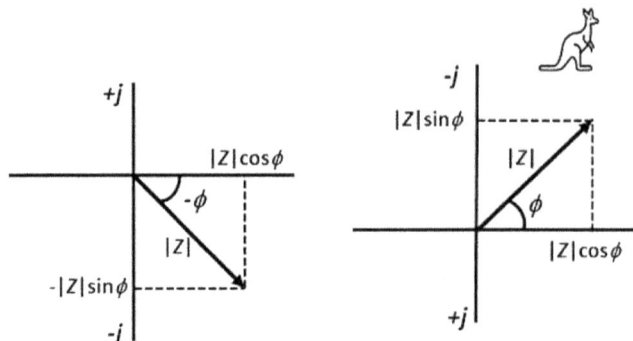

Figure 1.11. Representation of the impedance in the complex plane. The real part of the impedance is $|Z| \cos \phi$, and the imaginary part is $-|Z| \sin \phi$. The length of the arrow represents the magnitude of the impedance, $|Z|$. The right hand shows the conventional complex plane impedance plot with the $-j$ axis pointing up.

Figure 1.11 shows that we have two equivalent ways of representing an impedance. The first is to specify the magnitude of the impedance and the phase angle. The second is to specify the real and imaginary parts of the impedance. The two are of course related since

$$\tan \phi = \frac{\text{Im}[Z]}{\text{Re}[Z]} \quad \phi = \tan^{-1} \left(\frac{\text{Im}[Z]}{\text{Re}[Z]} \right) \tag{1.30}$$

From *Pythagoras theorem* for right-angled triangles, we discover that

$$|Z|^2 = \text{Re}[Z]^2 + \text{Im}[Z]^2 \tag{1.31a}$$

$$|Z| = \sqrt{\text{Re}[Z]^2 + \text{Im}[Z]^2} \tag{1.31b}$$

At the beginning of this chapter, we defined the admittance as the inverse of the impedance, i.e., $Y = 1/Z$. This makes sense – the more a circuit admits a current, the less it impedes it. However, we have just shown that the impedance is a complex quantity or a quantity with magnitude and phase. How do we take the inverse of a complex quantity? This is something we frequently need to be able to do in EIS, so we will deal with it in our next theory note.

Theory Note 1.4. Taking the Inverse of Complex Quantities
Suppose we have an impedance $Z = \text{Re}[Z] + j\text{Im}[Z]$. To find the admittance, Y, we need the inverse of Z, i.e.,

$$Y = \frac{1}{Z} = \frac{1}{\text{Re}[Z] + j\text{Im}[Z]} \tag{1.32}$$

Theory Note 1.4. (*Continued*)

The problem here is that whereas we can separate Z into real and imaginary components and plot it in the complex plane, we cannot do this for Y when it is expressed in this way. What we need is for Y to have the form $\text{Re}[Y] + j\text{Im}[Y]$. How do we achieve this? Here is the trick. We multiply the top and bottom of $1/Z$ by the *complex conjugate* of Z.

The complex conjugate of any complex number $x + jy$ is just $x - jy$, i.e., we just flip the sign of the imaginary part. So, let us see what happens if we multiply $1/Z$ by the complex conjugate of Z, which is $\text{Re}[Z] - j\text{Im}[Z]$. It will help to remember that $(a + b)(a - b) = a^2 + b^2$ and that $j^2 = -1$.

$$Y = \frac{1}{Z} = \frac{1}{\text{Re}[Z] + j\text{Im}[Z]} \times \frac{\text{Re}[Z] - j\text{Im}[Z]}{\text{Re}[Z] - j\text{Im}[Z]} = \frac{\text{Re}[Z] - j\text{Im}[Z]}{\text{Re}^2[Z] + \text{Im}^2[Z]} \quad (1.33)$$

Now we can separate Y into a real part and an imaginary part.

$$\text{Re}[Y] = \frac{\text{Re}[Z]}{\text{Re}^2[Z] + \text{Im}^2[Z]} \quad \text{Im}[Y] = \frac{-\text{Im}[Z]}{\text{Re}^2[Z] + \text{Im}^2[Z]} \quad (1.34)$$

It is then easy to show (*try it*) that the magnitude and phase angle of Y (Figure 1.12) are given by

$$|Y| = \frac{1}{|Z|} \quad \phi_Y = -\phi_Z \quad (1.35)$$

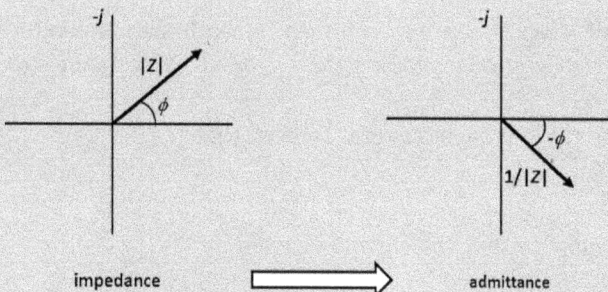

Figure 1.12. Inversion of impedance to obtain admittance. Note that the magnitude of Y is just the inverse of the magnitude of Z, and the phase angle simply changes sign going from Z to Y.

1.6 Linear Circuit Elements

Now let us return to the linear circuit elements R, C and L and look at their impedance. The resistor is easy. Its impedance is simply equal to the resistance. The applied voltage and current through an ideal resistor remain in phase at all frequencies, so there is no imaginary component, and therefore $Z = R$. The situation with capacitors and inductances is obviously more complicated, so we will deal with that now in our next theory note.

Theory Note 1.5. Impedances of Capacitors and Inductors

For a capacitor, we saw that

$$i = C\frac{dV}{dt} \tag{1.36}$$

Our ac voltage is a cosine waveform given by

$$V(t) = V_0 e^{j\omega t} \tag{1.37}$$

Here is where the complex exponential comes in use again. We need to differentiate $e^{j\omega t}$ with respect to t. Since differentiation of e^{ax} with respect to x gives ae^{ax},

$$\frac{dV}{dt} = V_0 j\omega e^{j\omega t} \quad i = C\frac{dV}{dt} = j\omega C V_0 e^{j\omega t} \tag{1.38}$$

Now we just need to remind ourselves that the impedance Z is the ratio $i(t)/V(t)$.

$$Z_C = \frac{V(t)}{i(t)} = \frac{V_0 e^{j\omega t}}{j\omega C V_0 e^{j\omega t}} = \frac{1}{j\omega C} = -\frac{j}{\omega C} \tag{1.39}$$

This result was obtained by multiplying both the numerator and the denominator by j, remembering that $j^2 = -1$. Note again that the $e^{j\omega t}$ terms cancel.

Now we will look at inductors. We saw that

$$V = L\frac{di}{dt} \tag{1.40}$$

We shall suppose that the current is given by

$$i(t) = i_0 e^{j\omega t} \tag{1.41}$$

Now we find the voltage by differentiating the $e^{j\omega t}$ term as before

$$V(t) = j\omega L \cdot i_0 e^{j\omega t} \tag{1.42}$$

The impedance of an inductor can now be found. It is

$$Z_L = \frac{V(t)}{i(t)} = \frac{j\omega L \cdot i_0 e^{j\omega t}}{i_0 e^{j\omega t}} = j\omega L \tag{1.43}$$

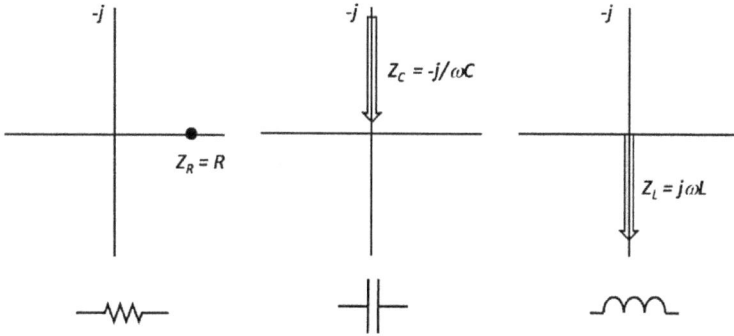

Figure 1.13. Complex plane plots of the three impedances – resistor, capacitor and inductor. The arrows show the direction of increasing frequency. The impedance of a resistor is constant, and the phase shift is zero. The impedance of a capacitor becomes infinite at zero frequency (dc), whereas the impedance of an inductor becomes zero at dc. Note that the y axis (imaginary axis) is $-j$. The phase shifts for capacitor and inductor are $-90°\,(-\pi/2)$ and $+90°\,(+\pi/2)$, respectively.

We now have three important expressions for the impedance of our three linear circuit elements.

$$Z_R = R \tag{1.44a}$$

$$Z_C = -\frac{j}{\omega C} \tag{1.44b}$$

$$Z_L = j\omega L \tag{1.44c}$$

Figure 1.13 shows each of these impedances plotted in the *complex plane*. Note that Z_C decreases as the frequency goes up, whereas Z_L goes the other way.

Now that we have defined the impedance of our three circuit elements, R, C and L, you could move on to the next chapter, which goes into the frequency-response analysis of RCL circuits in more detail. However, if you are interested in finding out more about the concept of *transfer functions* and how they determine the response of an electrical network to non-sinusoidal waveforms such as potential steps or ramps, then read on to find out more about the *complex frequency plane*, also called the *s-plane*, which is associated with the *Laplace transform*.

1.7 The Laplace Transform and Its Application in Circuit Analysis

The focus of this book is on frequency-response analysis of systems. However, there are many situations where we would like to understand

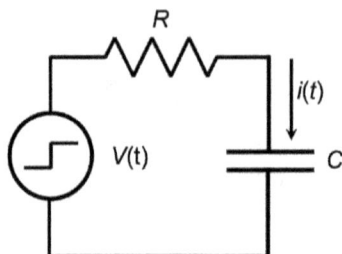

Figure 1.14.   Series circuit showing the voltage step $V(t) = V_A u(t)$ applied from the voltage source, where V_A is the amplitude and $u(t)$ is the *unit step function* (see following text for definition). The objective is to find the resulting time-dependence of the voltage across the capacitor $V_C(t)$ and of the current $i(t)$.

the response of systems to perturbations such as a sudden step in voltage or a steadily increasing voltage. We will begin with an example. Suppose that we suddenly switch on a voltage applied across a series connection of a resistor and a capacitor as shown in Figure 1.14.

Clearly, a current will flow when the voltage step is applied, and the capacitor will charge up as current flows through the resistor. Eventually, however, the voltage of the capacitor will reach the applied voltage and the current will fall to zero. How can we predict how the voltage builds across the capacitor and how the current falls? To tackle **time-dependent** problems of this kind we start by using *Kirchhoff's laws*. These are discussed in the next chapter (Section 2.3). First the current law. Since we have a series circuit, the current through the resistor and the capacitor must be the same. The current into the capacitor is given by

$$i(t) = C\frac{dV_c(t)}{dt} \tag{1.45}$$

where V_c is the voltage across the capacitor. The voltage across the resistor is given by Ohm's law, as follows:

$$V_R(t) = Ri(t) = RC\frac{dV_c(t)}{dt} \tag{1.46}$$

Now we use Kirchhoff's voltage law.

$$V(t) = V_R(t) + V_c(t) \tag{1.47}$$

which using equation (1.45) to define the current becomes

$$RC\frac{dV_c(t)}{dt} + V_c(t) = V(t) \tag{1.48}$$

The step function voltage $V(t)$ can be expressed as

$$V(t) = V_A u(t) \tag{1.49}$$

where V_A is the step height in volts and the *unit step function* $u(t)$ is written as follows:

$$u(t) = \begin{cases} 0 & t < 0 \\ 1 & t \geq 0 \end{cases} \tag{1.50}$$

The unit step function $u(t)$ is also known as the *Heaviside function*, named after the eccentric self-taught mathematician and physicist Oliver Heaviside, who is also known for the Kennelly–Heaviside layer of the ionosphere that makes it possible to bounce radio signal around the Earth. Our final *differential equation* in $V_C(t)$ is therefore

$$RC\frac{dV_c(t)}{dt} + V_c(t) = V_A u(t) \tag{1.51}$$

We would also like to find the time-dependent current in the circuit after applying the voltage step. We can guess that it decays exponentially as the capacitor charges up to the final voltage. To find out if this is correct, we need to use Kirchhoff's laws again. The voltage across the resistor is

$$V_R(t) = Ri(t) \tag{1.52}$$

The voltage across the capacitor at any time t depends on the charge $Q(t)$ that has flowed from $t = 0$, i.e., on the *integral* of the current (remember $Q = CV$),

$$V_c = \frac{1}{c} \int_0^t i(t)dt \tag{1.53}$$

From Kirchhoff's voltage law, we know that $V(t) = V_R + V_C$, so that

$$Ri(t) + \frac{1}{c} \int_0^t i(t)dt = V_A u(t) \tag{1.54}$$

How do we solve equations (1.51) and (1.54)? Although there are many ways of solving such differential equations, one of the most convenient is using *Laplace transforms*. The Laplace transform is named after Pierre Simon Laplace, the brilliant eighteenth-century mathematician and examiner of Napoleon at the École Militaire de Paris (Napoleon was ranked 42nd out of 58 in the final examination, which may tell us more about exams than about Napoleon). The transform takes a function $f(t)$ in the time domain and transforms it into the *complex frequency* or s *domain*. We already met the complex frequency domain in the context of periodic signals. In the Laplace

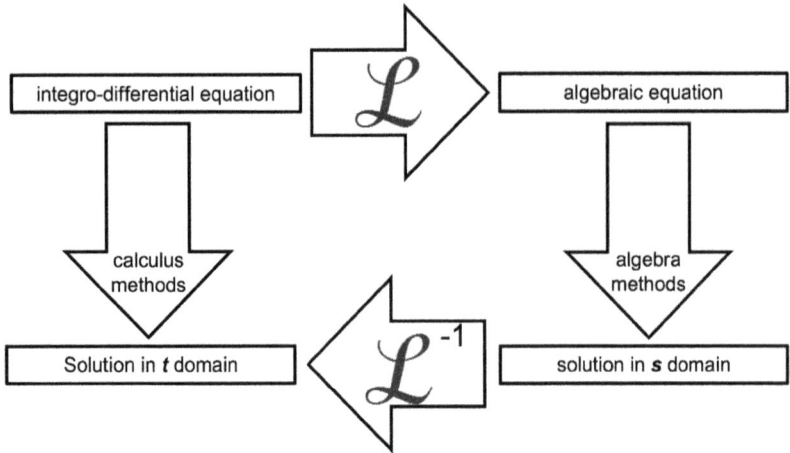

Figure 1.15. Comparison of routes to time-dependent solution of circuit responses. The Laplace transform and its inverse make it easier to obtain the solution by algebraic methods, avoiding calculus.

transformation, a more general definition of the *complex frequency variables s* allows us to obtain both transient and periodic responses. Although you might not think so, life in the complex frequency domain is simpler. We do not need to solve differential equations using calculus. Instead, we just need to do some relatively simple algebra to get the answer we want in the *s* domain. Once we have this solution, we perform an *inverse Laplace transform* to obtain the answer we want in the time domain. The basic idea is illustrated in Figure 1.15. Theory Note 1.6 takes a closer look at the Laplace transform.

Theory Note 1.6. A Short Introduction to the Laplace Transform

The Laplace transform of a function $f(t)$ is given by

$$\mathcal{L}f(t) = \mathcal{F}(s) = \int_0^\infty f(t)e^{-st}dt \tag{1.55}$$

where

$$s = \sigma + j\omega \tag{1.56}$$

is the *complex frequency variable*. The variable s can therefore be mapped on a plane with a real axis σ and an imaginary axis $j\omega$. This is the *s-plane* or *Laplace plane*.

Theory Note 1.6. (*Continued*)

Let us see what happens if $f(t)$ is the unit step function $u(t)$. For times greater than zero, $u(t)$ is just 1, so equation (1.55) becomes

$$F(s) = \int_0^\infty 1 \mathrm{e}^{-st} dt = \left[-\frac{1}{s}\mathrm{e}^{-st} \right]_0^\infty = 0 - \left(-\frac{1}{s} \right) = \frac{1}{s} \qquad (1.57)$$

The Laplace transform of the unit step function is therefore $1/s$.

Let us try another function. Suppose we have a voltage ramp that starts from zero and increases linearly with t. Our function $f(t)$ is now simply t and the Laplace transform becomes

$$F(s) = \int_0^\infty t \mathrm{e}^{-st} dt \qquad (1.58)$$

We need to *integrate by parts* using the formula that you may have learnt at school or college

$$\int u(t)v'(t) = u(t)v(t) - \int u'(t)v(t)dt \qquad (1.59)$$

Here, $v'(t)$ and $u'(t)$ are the derivatives of u and v with respect to time. So, we choose $u(t) = t$ and $v'(t) = \mathrm{e}^{-st}$, which means that $u' = 1$ and $v(t) = -\frac{1}{s}\mathrm{e}^{-st}$. Plugging these values into equation (1.59) gives

$$F(s) = \left[t \left(-\frac{1}{s}\mathrm{e}^{-st} \right) \right]_0^\infty + \int_0^\infty 1 \left(-\frac{1}{s}\mathrm{e}^{-st} \right) dt = 0 + \left[-\frac{1}{s^2}\mathrm{e}^{-st} \right]_0^\infty = \frac{1}{s^2} \qquad (1.60)$$

What about the Laplace transform of *derivatives* like those we saw when discussing the response of circuits to a voltage step – see equation (1.51)? If $f'(t)$ is the first derivative of $f(t)$, the Laplace transform is

$$\mathcal{L}f'(t) = \int_0^\infty f'(t)\mathrm{e}^{-st} dt \qquad (1.61)$$

Integrating by parts with $u = \mathrm{e}^{-st}$ and $v' = f'(t)dt$ gives

$$\mathcal{L}f'(t) = \left[\mathrm{e}^{-st} f(t) \right]_0^\infty + s \int_0^\infty f(t)\mathrm{e}^{-st} dt \qquad (1.62)$$

Evaluating the first term in equation (1.62) gives $-f(0)$, i.e., the negative of the value of the function $f(0)$ at $t = 0$. You should recognize the integral in the second term as $F(s)$, the Laplace transform of the function $f(t)$. The Laplace transform of the derivative $f'(t)$ is therefore given by

$$\mathcal{L}f'(t) = -f(0) + sF(s) = sF(s) - f(0) \qquad (1.63)$$

(*Continued*)

Theory Note 1.6. (*Continued*)

where $\mathcal{F}(s)$ is the Laplace transform of $f(t)$ and $f(0)$ is the value of $f(t)$ at $t = 0$.

A similar analysis gives the Laplace transform of an *integral* of a function – see equation (1.52). If $\mathcal{F}(s)$ is the Laplace transform of $f(t)$, then

$$\mathcal{L} \int_0^t f(t)dt = \frac{\mathcal{F}(s)}{s} \tag{1.64}$$

You can find tables of Laplace transforms in Maths textbooks or on the web. Table 1.2 gives some of the most useful Laplace transforms for the analysis of RCL circuits. The table includes transforms of functions that are derivatives or integrals, and these will be very useful for problems like the RC or RL circuit discussed previously.

Table 1.2. Laplace transforms of some common time-dependent functions.

Function type	$f(t)$	$F(s) = \mathcal{L}f(t)$
Unit step	$f(t) = 1$ for $t > 0$	S
Exponential decay	$f(t) = e^{-at}$	$\frac{1}{s}$
Ramp	$f(t) = t$	$\frac{1}{s^2}$
Sine	$f(t) = \sin(\beta t)$	$\frac{\beta}{s^2 + \beta^2}$
Cosine	$f(t) = \cos(\alpha t)$	$\frac{s}{s^2 + \alpha^2}$
First derivative	$f'(t)$	$s\mathcal{F}(s) - f(0)$
Integral	$\int_0^t f(t)dt$	$\frac{\mathcal{F}(s)}{s}$

We could look at more examples, but the takeaway message is that the Laplace transform gives expressions containing the complex frequency variable s that can be manipulated with *basic algebra*. The inverse Laplace transform takes these simple expressions and converts them back to time-domain functions as shown in Figure 1.15.

We can now apply the Laplace transform method to calculate the charging of the capacitor in the circuit in Figure 1.14 when we apply a voltage step. The next theory note shows how we do this.

Theory Note 1.7. Using the Laplace Transform to Obtain Time-Dependent Solutions

For the RC circuit in Figure 1.14, we obtained the following differential equation for the voltage across the capacitor:

$$RC\frac{dV_c(t)}{dt} + V_c(t) = V(t) \tag{1.65}$$

Here, the applied voltage $V(t)$ is a voltage step $V_A u(t)$, where $u(t)$ is the unit step function. Our objective is to solve this equation to find $V_c(t)$. This means that we want to obtain $\hat{V}_c(s)$ and then use the inverse Laplace transform to obtain $V_c(t)$. The $\wedge$ symbol over voltage and currents is used to indicate that we are working in the Laplace plane. We start by applying the Laplace transform to each term in the equation, as follows:

$$\mathcal{L}\left[RC\frac{dV_c(t)}{dt}\right] + \mathcal{L}\left[V_c(t)\right] = \mathcal{L}\left[V_A u(t)\right] \tag{1.66}$$

The first term is the Laplace transform of a derivative (RC is a constant). The second term is the Laplace transform of the quantity we want to find. The third term is the Laplace transform of the unit step function (V_A is a constant). Referring to Table 1.2, we find that equation (1.64) becomes

$$RC\left[s\hat{V}_c(s) - V_c(0)\right] + \hat{V}_c(s) = \frac{V_A}{s} \tag{1.67}$$

We will assume that the voltage across the capacitor is initially zero ($V_C = 0$). Collecting up terms in $\hat{V}_c(s)$ gives

$$\hat{V}_c(s) = \frac{V_A}{s(1 + sRC)} \tag{1.68}$$

Looking ahead to the inverse Laplace transform we want terms like $(s + a)$, where a is a constant, so we divide the numerator and denominator of the right-hand side of equation (1.68) by RC to obtain

$$\hat{V}_c(s) = \frac{V_A}{RC}\frac{1}{s\left(s + \frac{1}{RC}\right)} \tag{1.69}$$

Here the constant a is $1/RC$.

Our next task is to use *partial fractions* to break the fraction on the right-hand side of equation (1.69) into the sum of two fractions. We do this as follows. We let

$$\frac{V_A}{RC}\frac{1}{s\left(s + \frac{1}{RC}\right)} = \left[\frac{A}{s} + \frac{B}{\left(s + \frac{1}{RC}\right)}\right] \tag{1.70}$$

(Continued)

Theory Note 1.7. (*Continued*)

where A and B need to be found. Multiplying both sides by $s\left(s + \frac{1}{RC}\right)$ gives

$$\frac{V_A}{RC} = A\left(s + \frac{1}{RC}\right) + Bs \qquad (1.71)$$

which we can rearrange to

$$A\left(s + \frac{1}{RC}\right) + Bs - \frac{V_A}{RC} = 0 \qquad (1.72)$$

Multiplying out and collecting the terms in s and constant terms gives

$$(A + B)s + \left(\frac{A - V_A}{RC}\right) = 0 \qquad (1.73)$$

It follows that both terms in equation (1.73) must be zero, we find that $A = V_A$ and $B = -V_A$. We can now rewrite equation (1.69) as

$$\hat{V}_c(s) = \left[\frac{V_A}{s} - \frac{V_A}{\left(s + \frac{1}{RC}\right)}\right] \qquad (1.74)$$

Now we are ready to use the inverse transform to obtain $V_c(t)$. Inverse transformation of the first term in the square brackets gives $V_A u(t)$. If you check Table 1.2, you will see that the second term transforms to $-V_A e^{-\left(\frac{t}{RC}\right)}$, so that

$$V_c = V_A\left[1 - e^{-\left(\frac{t}{RC}\right)}\right] \qquad (1.75)$$

Figure 1.16 shows how the capacitor charges up to the final voltage V_A. By the time $t = 4RC$, the capacitor is nearing full charge. The product RC is of course the *time constant* of the series RC combination.

Now we can move on to tackle the problem of finding the current transient $i(t)$ for the charging of the capacitor in Figure 1.14. We start by taking the Laplace transforms of the terms in equation (1.54). We obtain

$$Ri\mathcal{L}(t) + \frac{1}{c}\mathcal{L}\int_0^t i(t)dt = V_A\mathcal{L}u(t) = R\hat{i}(s) + \frac{1}{Cs}\hat{i}(s) = \frac{V_A}{s} \qquad (1.76)$$

from which we obtain $\hat{i}(s)$, the Laplace transform of $i(t)$ as follows:

$$\hat{i}(s) = \frac{V_A}{R}\left(\frac{1}{s + \frac{1}{RC}}\right) \qquad (1.77)$$

Checking the table of Laplace transforms, we recognize the term in brackets as having the form $\frac{1}{(s+a)}$, which is the Laplace transform of

Theory Note 1.7. (*Continued*)

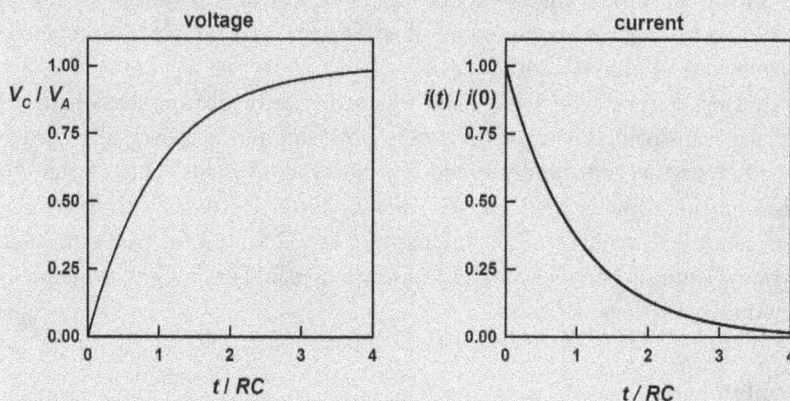

Figure 1.16. The normalized plot on the left shows how the voltage across the capacitor approaches the applied voltage as it is charged through the series resistor. The right-hand plot shows the exponential decay of the charging current, which has been normalized by dividing by the initial charging current $i(0) = \frac{V_A}{R}$.

a decaying exponential. Our final expression for the decaying charging current is therefore

$$i(t) = \frac{V_A}{R}e^{-\left(\frac{t}{RC}\right)} \tag{1.78}$$

Figure 1.16 shows the time-dependence of the voltage across the capacitor and the current in the circuit.

If we consider any two-terminal circuit containing RCL elements, we can define the Laplace transforms of the current in the circuit, $\mathcal{L}i(t) = \hat{i}(s)$, and of the voltage across the circuit, $\mathcal{L}V(t) = \hat{V}(s)$. This allows us to define $\hat{Z}(s)$, the *operational impedance* in the s (complex frequency) domain as

$$\hat{Z}(s) = \frac{\hat{V}(s)}{\hat{i}(s)} \tag{1.79}$$

Now we can obtain the operational impedance of simple resistors, capacitors and inductors. It is easy to show that

$$\hat{Z}_R = R \tag{1.80a}$$

$$\hat{Z}_C = \frac{1}{sC} \tag{1.80b}$$

$$\hat{Z}_L = sL \tag{1.80c}$$

Remembering that $s = \sigma + j\omega$, we can see that these expressions are the same as those that we found in Section 1.6 – equations (1.44a)–(1.44c) if σ, the real component of s, is zero so that we can replace s in equations (1.80a)–(1.80c) by $j\omega$.

In this chapter, we have only touched briefly on application of the Laplace transform. The method is very powerful and is universally used by electrical engineers to analyze signal processing systems. One of the best books on this topic is *The Scientist and Engineer's Guide to Digital Signal Processing* by Steven W. Smith. Chapters 30–32 are useful background for anyone wishing to understand EIS in more depth. The book is available on the Analog Devices website.

Problems

(Answers to the problems are given at the end of the book.)

1.1. The impedance of an RC circuit at a particular frequency is given in polar coordinates by $|Z| = 1414\Omega, \phi = -\pi/4$. Plot this impedance as a vector in the complex plane. Express the impedance in the form $Z = \text{Re}[Z] + j\text{Im}[Z]$.

1.2. Calculate the admittance from the impedance in problem 1.1. Express the answer in polar coordinates $|Y|, \theta$ and in the form $Y = \text{Re}[Y] + j\text{Im}[Y]$. Plot the admittance as a vector in the complex plane.

1.3. Calculate the impedance of a 10^{-7} F capacitor at frequencies 1, 10, 100 and 1000 Hz. Make the same calculation for a 10^{-3} H inductor. Sketch Bode graphs to contrast the frequency dependence of the impedance in the two cases.

1.4. At what frequency (in Hz) will the impedance of a 10^{-7} F capacitor and a 10^{-3} H inductor have exactly the same magnitude? Do not try to obtain this from your graphs in problem 1.3. Instead, try to derive an expression for the 'magic frequency'.

1.5. Consider the following circuit. Can you predict the shape of the current transient when a single step voltage of amplitude 1 V is applied? Sketch the shape of the current transient. Explain the initial and final current values and any other features, such as current decay. You do not need to solve the problem by Laplace transforms – just use 'common sense'.

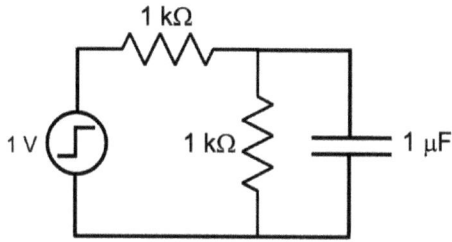

1.6. Consider the following circuit. Can you predict the shape of the current transient when a single step voltage of amplitude 1 V is applied? Sketch the shape of the current transient. Explain the initial and final current values and any other features, such as current decay. Again, you do not need to solve the problem by Laplace transforms – just use 'common sense'. Hint: You will need to think about the units of inductance when defining a time constant.

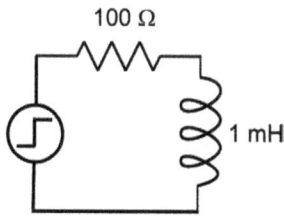

Chapter 2

Frequency-Response Analysis

2.1 Modelling Impedance Responses Using ZView®

This is a 'nuts and bolts' chapter. It looks at how the impedance of different electrical circuits changes with the frequency of the ac signal. We introduce the ZView® impedance simulation and fitting program and use it to produce characteristic complex plane plots (*Nyquist plots*) and *Bode plots* of impedance and admittance for a wide range of simple RC circuits.

ZView® is probably the most widely used impedance simulation and nonlinear least squares fitting software program for electrochemical impedance spectroscopy (EIS). It is available from *Scribner Associates*, a company that grew out of the Center for Electrochemical Science and Engineering at the University of Virginia. I am grateful to Scribner for permission to use ZView® to produce some of the figures in this book.

In this chapter, ZView® is used in simulation mode to create plots of impedance as a function of frequency. We can create any circuit in ZView® and then enter values for the linear circuit elements, e.g., two resistors and a capacitance as in the example shown in Figure 2.1. We then click on 'model' and chose the *simulation mode* in the window shown in Figure 2.2 to run the calculation over the default frequency range 0.001 Hz to 10^6 Hz.

The calculated impedance data can be displayed in two ways. The first is to plot the real and imaginary components of the impedance in the *complex plane*, which we encountered in Chapter 1. This plot is the *Nyquist plot*, named after the Swedish-American electrical engineer Harry Nyquist from Bell Telephone Laboratories, who introduced the concept of complex plane plots to represent the transfer functions of amplifiers (see Chapter 1) and

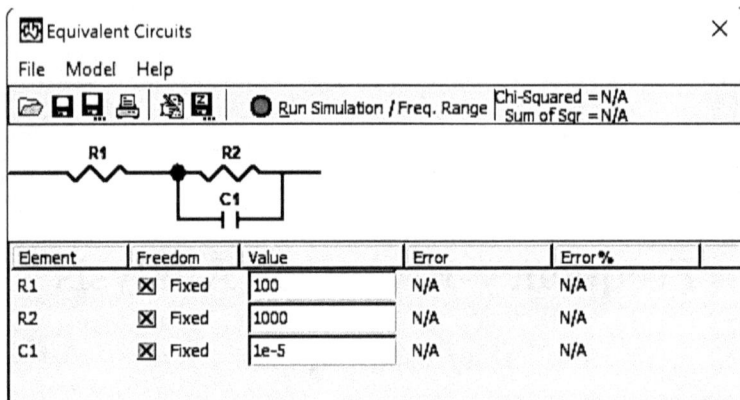

Figure 2.1. Inputting linear circuit elements into the ZView® Equivalent Circuits model.

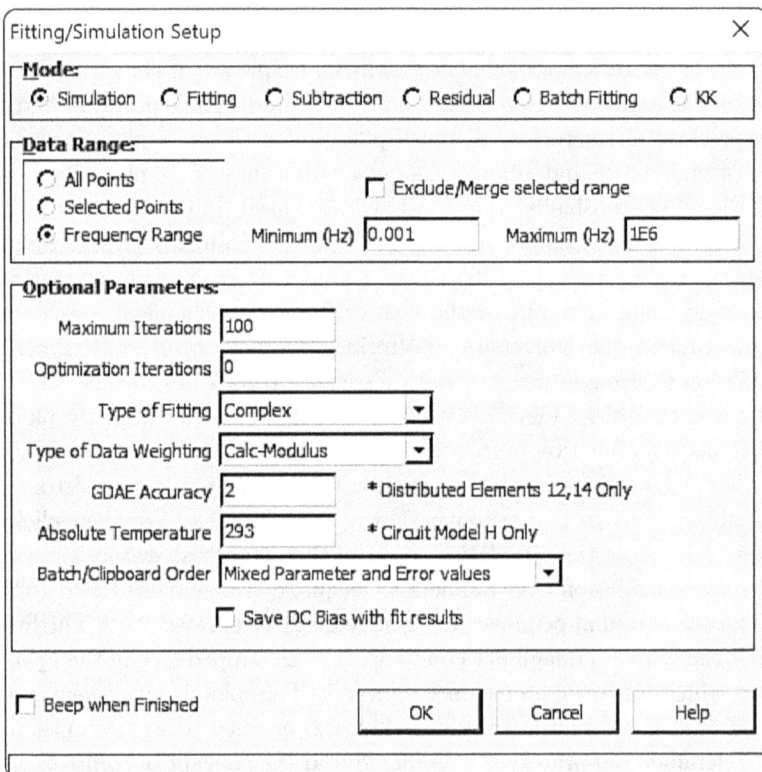

Figure 2.2. Running ZView® in simulation mode to calculate the frequency response for the circuit in Figure 2.1.

to define whether they will be stable or unstable. The term has come to be more widely used for other complex plane plots of other transfer functions including impedance and admittance. The second way of plotting the data is the *Bode plot*, introduced by Hendrik Wade Bode (also from the Bell Telephone Laboratories) in the 1930s to represent the gain and phase shift of amplifiers over a wide frequency range. Again, the term has been adopted for impedance data. There are two Bode plots for impedance data – one of $\log_{10}|Z|$ vs. $\log_{10}f$, the other of phase shift vs. $\log_{10}f$. Sometimes the real and imaginary components of Z are plotted against $\log_{10}f$. Figure 2.3 shows the Nyquist plot of the impedance of the circuit in Figure 2.1, and Figure 2.4 shows the corresponding Bode plots of magnitude and phase (note the logarithmic scales for Z and f). Note that unless the points are labelled, the Nyquist plot provides no frequency information. It is therefore good practice to label the plots with relevant frequencies.

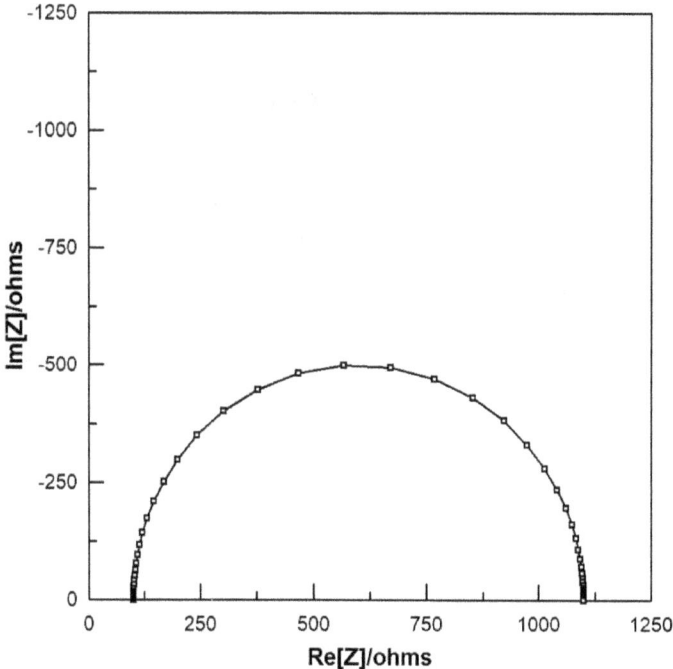

Figure 2.3. Nyquist plot of impedance data calculated by ZView® for the circuit in Figure 2.1. When presenting experimental results, it is good practice to label several frequency values on Nyquist plots.

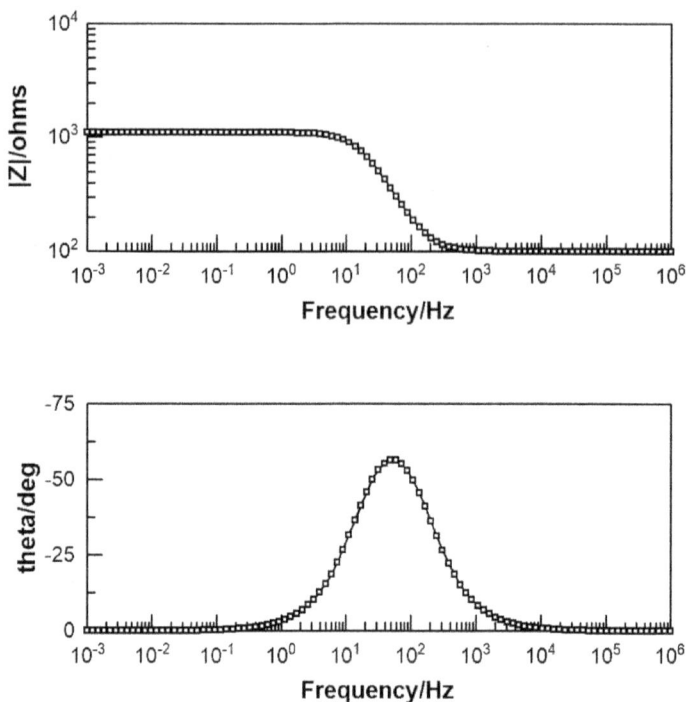

Figure 2.4.   The Bode plots of magnitude and phase for the circuit in Figure 2.1.

2.2 The Impedance of Resistors, Capacitors, and Inductors

We will return to the circuit shown in Figure 2.1 once we have looked at simpler circuits. We begin by looking at the impedance responses of the single linear elements R, C and L – resistor, capacitor and inductor. Before we look at the impedance plots, here are the impedance–frequency relationships for all three elements, R, C and L, that we saw in Chapter 1.

The three impedance equations

$$Z_R = R \tag{2.1}$$

$$Z_C = -\frac{j}{\omega C} = -\frac{j}{2\pi f C} \tag{2.2}$$

$$Z_L = j\omega L = j2\pi f L \tag{2.3}$$

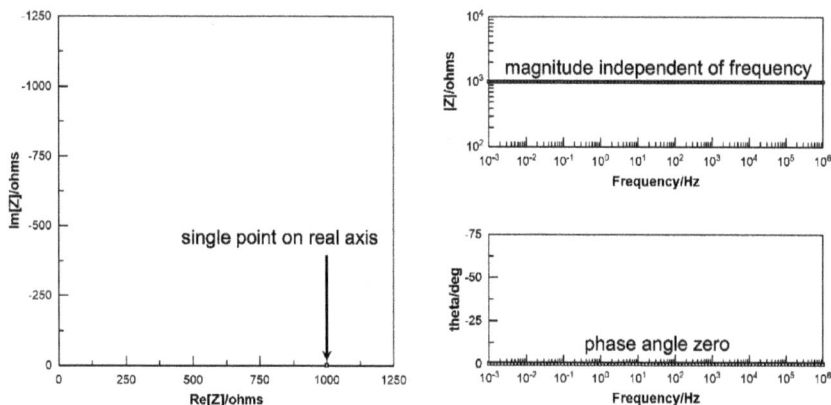

Figure 2.5. Nyquist and Bode plots for a resistor (1000 Ω) – not very exciting.

As Figure 2.5 shows, the impedance response of a resistor (1000 Ω) is a *single point on the real axis* in the Nyquist plot. The Bode plot of $\log_{10}|Z|$ vs. $\log_{10}f$ is a horizontal straight line, and the phase angle plot shows a constant phase shift of 0°.

Now for the capacitor, C. Here we know that the impedance decreases as the frequency increases. We also know that the impedance is entirely imaginary ($-j$). This means that the phase shift is always $-90°$. The Nyquist and Bode plots are shown in Figure 2.6.

Which brings us to the inductor, L. In this case, the impedance increases with frequency. No prizes for guessing what the impedance plots look like – see Figure 2.7. Now as the frequency increases from zero to higher values, the Nyquist plot for the inductor runs from the origin down the y axis (note that we are still using the 'Australian' convention introduced in Chapter 1, i.e., negative imaginary components are up, and positive are down). The Bode plot of $\log_{10}|Z|$ vs. $\log_{10}f$ now has a *positive* slope of 1 because the impedance is a linear function of frequency, i.e., Z_L depends on $f^{+1.0}$ (see Theory Note 2.1). We see that the capacitor and the inductor respond in opposite ways when the frequency is increased. This raises an interesting question. What happens if we have a capacitor and an inductor connected *in series*, i.e., one after the other? This question will be addressed in the next section, where we begin to build up *RCL* networks.

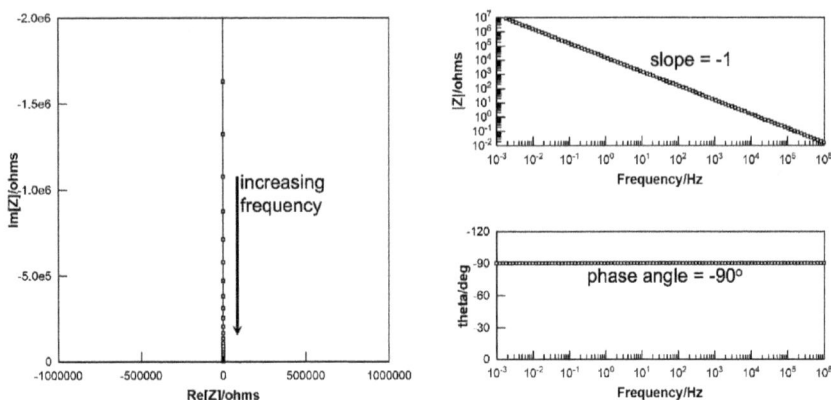

Figure 2.6. Nyquist and Bode plots for a capacitor $(10^{-5}\,\text{F} = 10\,\mu\text{F})$. Slightly more interesting. Note that the impedance is infinite for zero frequency (dc) and tends towards zero as the frequency tends towards infinity. The slope of the $\log_{10}|Z|$ vs. $\log_{10}f$ plot is -1. The phase angle is always $-90°$.

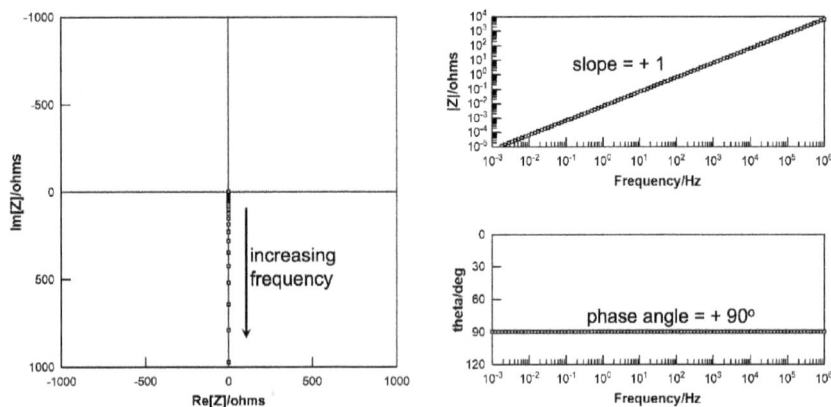

Figure 2.7. Nyquist and Bode plots for an inductor $(10^{-3}\,\text{H or 1 mH})$. Note that *the imaginary component is positive* (it is negative for a capacitor). The phase angle is a constant $+90°$.

Theory Note 2.1. Why the Slope of the $\log_{10}|Z|$ vs. $\log_{10}f$ Bode Plot in Figure 2.6 Is -1

We know that the impedance of a capacitor is given by $Z_C = -\frac{j}{2\pi fC}$. Taking logarithms to base 10 (i.e., $\log_{10}$) of both sides of this equation, we obtain

$$\log_{10}Z_C = \log_{10}(-j) - \log_{10}(2\pi fC) = \log_{10}(-j) - \log_{10}(2\pi C) - \log_{10}f \tag{2.4}$$

Here, we used the properties of logarithms: $\log_{10}(a/b) = \log_{10}a - \log_{10}b$ and $\log_{10}cd = \log_{10}c + \log_{10}d$.

On the right-hand side of equation (2.4), everything is constant except the $\log_{10}f$ term, so we can write

$$\log_{10}Z_C = \text{constant} - \log_{10}f = \text{constant} + (-1)\log_{10}f \tag{2.5}$$

which is the *equation for a straight line* in the form $y = c + mx$, where c is a constant and m is the slope of a plot of y against x (in our case, $m = -1$). We see why the slope of the Bode plot is -1. It is simply because the impedance is *inversely proportional* to the frequency. In the case of an inductor, Z_L depends *linearly on frequency*, so the slope of the $\log_{10}|Z_L|$ vs. $\log f$ Bode plot will have a slope of $+1$.

2.3 Kirchhoff's Laws

To tackle RCL networks, we need some basic rules and some common sense. Let us start with the rules. These are summarized by Kirchhoff's laws or, more correctly, Kirchhoff's *circuit* laws, formulated in 1845 by the German physicist Gustav Kirchhoff (Kirchhoff also formulated laws in spectroscopy, thermal radiation and thermochemistry). You probably came across the circuit laws already at school or college. There are two Kirchhoff laws. The first deals with currents. And it may seem obvious. Basically, it says that if several wires connect at a point or *node*, what goes in must come out. This can be expressed more elegantly by defining currents flowing into a node as positive and current flowing out of a node as negative. *Kirchhoff's current law* says that for any node the *algebraic sum of the currents* (i.e., taking their magnitude and sign) *is zero*. This simply means that currents cannot appear from nowhere or disappear into nowhere. The second law deals with voltages. These can be voltages across resistors, capacitors or inductors, or they can be voltages created, for example, by a battery in a circuit. *Kirchhoff's voltage law* says that *the sum of the voltages going*

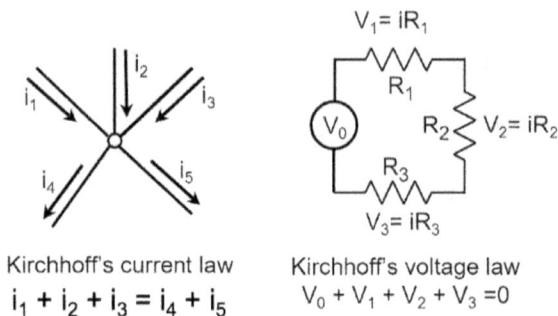

Figure 2.8. Kirchhoff's laws. The current law states that the algebraic sum of currents flowing into a node is zero. The currents i_1, i_2 and i_3 have a positive sign. The currents i_4 and i_5 have a negative sign. The voltage law states that if we go around a loop in one direction, the sum of the voltages is equal to zero. In the figure, V_0, the voltage of the voltage source (e.g., battery), is equal and opposite to the sum of the voltage drops across the three resistors in the loop.

around a closed loop in one direction *is zero*. The two laws are illustrated in Figure 2.8.

2.4 *RCL* Networks

From Kirchhoff's laws, one can derive the following two important relationships for impedances connected in *series* or in *parallel*. For a series connection, simply add the impedances. For a parallel connection, add the inverses of the impedances (i.e., add the admittances) to obtain the inverse of the total impedance. The procedure is illustrated in Figure 2.9.

In a case where the circuit contains both series and parallel connections like the one in Figure 2.10, the impedances are calculated in a stepwise sequence as shown.

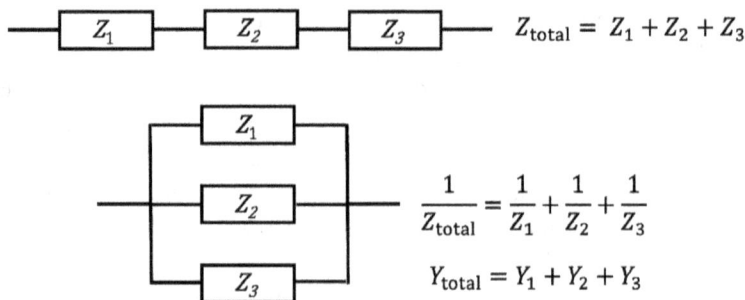

Figure 2.9. Working out the total impedance and admittance for series and parallel connections.

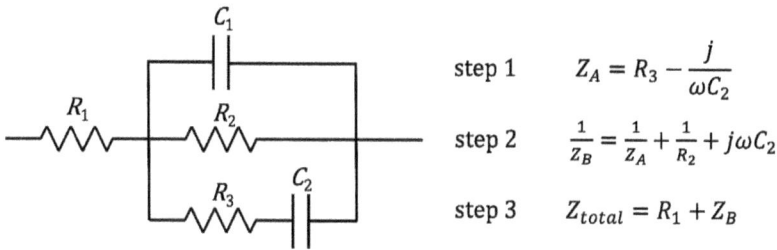

step 1 $\quad Z_A = R_3 - \dfrac{j}{\omega C_2}$

step 2 $\quad \dfrac{1}{Z_B} = \dfrac{1}{Z_A} + \dfrac{1}{R_2} + j\omega C_2$

step 3 $\quad Z_{total} = R_1 + Z_B$

Figure 2.10. An example of the steps used when working out the impedance of series/parallel circuits.

In principle, you can now work out an expression for the total impedance of the circuit in Figure 2.10. The procedure follows the steps shown above and involves using the *complex conjugate* wherever necessary to obtain the real and imaginary components. The derivation is given in Appendix 2.1 together with an illustrative calculation for given values of the resistors and capacitors. This appendix is worth studying in some detail.

2.5 Series Connections

Now we are equipped to tackle quite complicated networks of linear circuit elements. We start with the one mentioned in the previous section, the series connection of a resistor and an inductor (Figure 2.11).

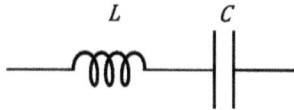

Figure 2.11. Series connection of a capacitor and an inductor. In the calculated example shown in Figure 2.12, $C = 10^{-5}$ F ($10\,\mu$F) and $L = 10^{-3}$ H (1 mH).

Before we use ZView® to calculate the impedance response of this circuit, we will apply some common sense. It is useful to ask two simple questions.

(Q1) What will the impedance of the circuit be for *zero frequency* (i.e., for dc)?

(Q2) What will the impedance of the circuit be for *infinite frequency*?

The impedance of the capacitor at zero frequency is infinite – it cannot pass a dc current. The impedance of the inductor is zero for dc. So, the answer to the first question is that the impedance for $f = 0$ is $-j\infty + 0 = -j\infty$ (note the negative sign). What about the second question? Well, the impedance

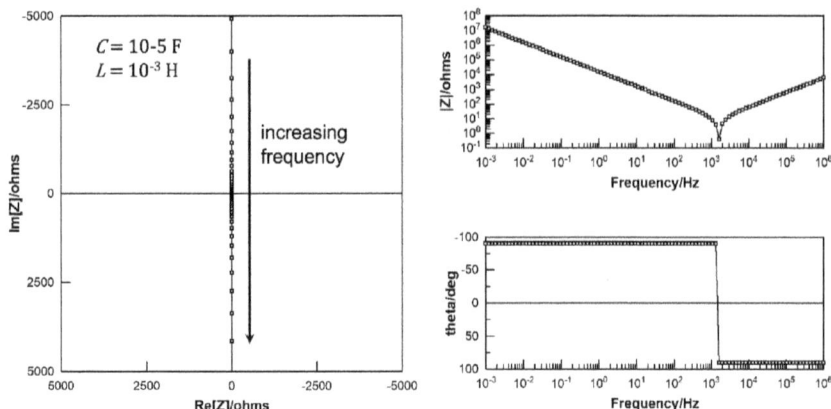

Figure 2.12. Impedance response of a capacitor and an inductor in series. $C = 10^{-5}$ F, $L = 10^{-3}$ H.

of the capacitor at infinite frequency will be zero, but now it is the turn of the inductor to have an infinite impedance. So, the answer to Question 2 is that the impedance for $f = \infty$ is $0 + j\infty = j\infty$ (note the positive sign). We now have the low-frequency and high-frequency limiting cases for the impedance – both are imaginary and infinite, but with opposite signs. The Nyquist plot of the impedance must therefore cross from the $-j$ side to the $+j$ side as the frequency increases, so on the way *it must go through zero*.

Now, let us turn to ZView® to find out what this means. The ZView® plot confirms our common-sense result. At low frequencies, the Nyquist plot looks like the plot we saw earlier for a capacitor, i.e., it comes down the $-j$ axis towards the origin. However, instead of tending towards zero at high frequencies, *the plot passes through zero* before heading off down the $+j$ axis. Now look at the two Bode plots. At low frequencies, the impedance decreases with increasing frequency and the phase angle is $-90°$, i.e., the circuit behaves like a capacitor. At high frequencies, on the other hand, the impedance increases again, and the phase angle is $+90°$, i.e., our circuit looks like a simple inductor. The interesting thing is that between these two limits and *at a particular frequency, the impedance vanishes*, and the phase angle switches abruptly from $-90°$ to $+90°$. At this point, our circuit just looks like a zero-resistance wire!

So, what determines the unique frequency (let us call it ω_0 in terms of radial frequency) at which the impedance becomes zero? It must be true that the impedances of the capacitor and inductor are *equal and opposite* at the frequency ω_0. Let us look at what that means.

$$Z_C + Z_L = 0$$

$$-\frac{j}{\omega_0 C} + j\omega_0 L = 0$$

$$\omega_0 = 2\pi f_0 = \frac{1}{\sqrt{LC}} \quad \text{and} \quad f_0 = \frac{1}{2\pi\sqrt{LC}} \tag{2.6}$$

We see that the unique frequency, f_0, where the impedance is zero is determined by the inverse of the square root of the product LC. This means we can *tune* f_0 by changing L or C. Here, $C = 10^{-5}$ F and $L = 10^{-3}$ H, giving $f_0 = 1.59$ kHz, which is exactly the frequency at which the phase angle flips from negative to positive in the Bode plot and the impedance goes through zero. This is called the *resonant frequency* of the circuit. Note that the units of L are $\Omega s = VA^{-1} s$ and those of C are $C V^{-1} = A s V^{-1}$. The product LC therefore has units s^2, and $\sqrt{LC}$ has units of s, i.e., it is a time constant.

In EIS, we mostly consider inductance to be a nuisance (the wires used for connecting to a potentiostat or frequency response analyzer (FRA) have a *self-inductance* of the order of 10^{-6} H/m, so it is important to keep connections as short as possible for high-frequency EIS measurements). Initially, we will focus just on resistors and capacitors. Let us put a resistor and a capacitor in series (Figure 2.13).

Now we must ask two simple questions:

(Q1) What will the impedance of the circuit be for *zero frequency* (i.e., for dc)?

(Q2) What will the impedance of the circuit be for *infinite frequency*?

For zero frequency, as in the LC example, the impedance is infinite. But now when the frequency becomes large, the impedance of the capacitor becomes very small, so we are left just with the resistor, i.e., the impedance will tend towards R, which we saw is just a point on the real axis. So, we expect our Nyquist plot to look like the one for a capacitor (look back at Figure 2.6

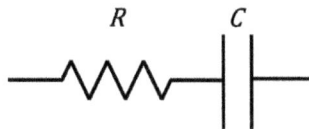

Figure 2.13. Series RC circuit. In the following example, $R = 10^3 \Omega (1 \text{ k}\Omega)$, $C = 10^{-5}$ F $(1 \mu\text{F})$.

to remind yourself what this looked like). The difference is that instead of tending towards the origin $(0, 0)$ at high frequency, the vertical Nyquist plot should hit the real axis at the value of the resistor R. The ZView® result in Figure 2.14 confirms this. It is also clear on the Bode plot, where the $\log_{10}|Z|$ vs. $\log_{10}f$ plot flattens off at $10^3\,\Omega$ at high frequencies.

By now you should be getting used to using clues in the Bode plot to understand the behaviour of circuits. In this case, the slope of -1 in the $\log |\log_{10}|Z|$ vs. $\log_{10}f$ plot and the phase angle of $-90°$ tells us we have a capacitor that dominates the low-frequency response. At high frequencies, $|Z|$ becomes constant and the phase angle becomes $0°$. Clearly, we have a series resistor that dominates the high-frequency response.

We saw there was a magical resonance frequency for our series LC circuit. What about the series RC circuit? Again, let us consider the frequency ω_0 where the magnitudes of the impedance of the resistor and capacitor $|Z_C|$ and $|Z_R|$ are equal.

$$|Z_C| = \sqrt{\mathrm{Re}^2 Z_C + \mathrm{Im}^2 Z_c} = \sqrt{0 + \frac{1}{(\omega_0 C)^2}} = \frac{1}{\omega_0 C} \quad \text{and } |Z_R| = R$$

$$(2.7a)$$

If

$$|Z_R| = |Z_C|, R = \frac{1}{\omega_0 C} \quad \text{and} \quad \omega_0 = \frac{1}{RC} \; f_0 = \frac{1}{2\pi RC} \qquad (2.7b)$$

Putting the values for R ($10^3\Omega$) and C ($10^{-6}\,\mathrm{F}$) into equation (2.7b), we find that $f_0 = 15.9\,\mathrm{Hz}$ ($\omega_0 = 100\,\mathrm{rad\,s^{-1}}$). Let us look at what ZView® showed us. The transition from a slope of -1 to a horizontal line in the Bode plot occurs at this frequency. Also, the phase angle at this frequency

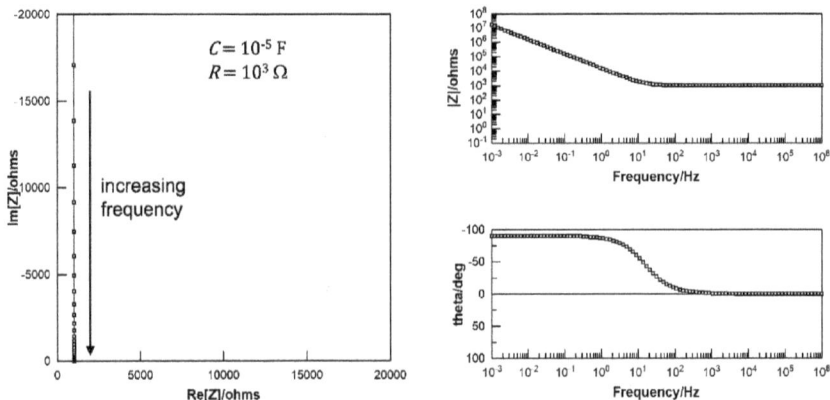

Figure 2.14. Nyquist and Bode plots of the impedance response of a series RC circuit. $R = 10^3\,\Omega, C = 10^{-5}\,\mathrm{F}$.

is halfway between $-90°$ and zero, i.e., it is $-45°(-\pi/4)$. Time for a Theory Note to explain what is going on here.

Theory Note 2.2. Impedance of a Series RC Circuit

The impedance of two impedance elements in series is simply $Z_{\text{total}} = Z_1 + Z_2$. In the case of our series RC circuit,

$$Z_{\text{total}} = Z_R + Z_C = R - \frac{j}{\omega C} \qquad (2.8)$$

We see at once that the real part of Z_{total} is just R, and the imaginary part is $-1/\omega C$.

The magnitude of Z_{total} is therefore

$$|Z_{\text{total}}| = \sqrt{\text{Re}^2 Z_{\text{total}} + \text{Im}^2 Z_{\text{total}}} = \sqrt{R^2 + \frac{1}{\omega^2 C^2}} \qquad (2.9)$$

The phase angle is given by

$$\tan^{-1}\theta = \frac{\text{Im}\,[Z_{\text{tot}}]}{\text{Re}\,[Z_{\text{tot}}]} = \frac{-1}{\omega RC} \qquad (2.10)$$

We are interested in the special case where $|Z_R| = |Z_C|$, i.e., $R = 1/\omega C$. This means that equation (2.9) tells us that the magnitude of the impedance $|Z_{\text{total}}|$ in this case is equal to $\sqrt{2R^2} = \sqrt{2}R$. Equation (2.10) simplifies to $\tan^{-1}\theta = -1$ if $R = 1/\omega C$, so the phase angle must be $-45°(-\pi/4)$.

We can have a look at this using a complex plane plot of the impedance. We know that the real and imaginary components of Z_{total} are both equal to $\sqrt{2}R$ at the 'magic frequency' $\omega_0 = 2\pi f_0$, so the plot in Figure 2.15 makes sense of what we just worked out.

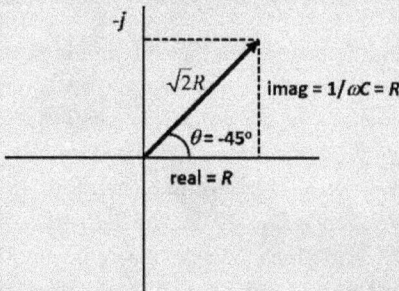

Figure 2.15. Impedance of the series RC circuit at the frequency $\omega_0 = 2\pi f_0 = 1/RC$ radians per second. Note that since the real and imaginary components are equal, the magnitude of the impedance must be $\sqrt{2}R$ and the phase angle must be $-45°(-\pi/4)$. Note the minus sign arises from the fact that the imaginary component is negative, i.e., the upward y axis is the $-j$ axis.

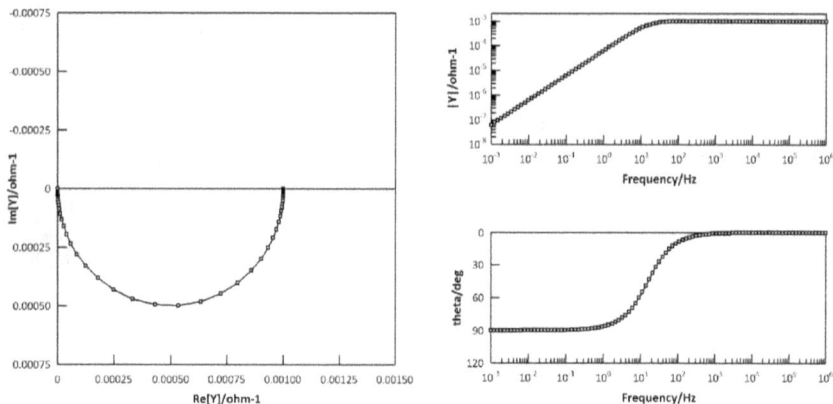

Figure 2.16. Admittance plots for the series RC circuit ($R = 10^3\,\Omega$, $C = 10^{-5}\,\mathrm{F}$). Note that the admittance is a semicircle going from the origin at zero frequency to $1/R$ at high frequency.

So far, we have only looked at impedance. What about *admittance*? Since ZView® allows us to display the results as admittance, let us have a look at the admittance of the series RC circuit. The result is interesting. The admittance is a semicircle going from the origin at zero frequency to $1/R$ at high frequency (Figure 2.16).

Since the semicircle appears 'upside down', the imaginary component of the admittance must be positive. Can we rationalize the admittance plot using our 'common-sense' approach of asking two simple questions?

(1) What will the admittance of the circuit be for *zero frequency* (i.e., for dc)?
(2) What will the admittance of the circuit be for *infinite frequency*?

The capacitor will block dc, so *the admittance will be zero at zero frequency*. At infinite frequency, the admittance will be due only to the resistor (the capacitor will effectively be a short circuit), therefore, *the admittance at high frequencies will be 1/R*. Since $R = 10^3\Omega$, the high-frequency admittance will be $1/10^3 = 10^{-3}\Omega^{-1}$, and since R is just a real quantity, the admittance point will also be on the real axis. 0 and $1/R$ are clearly the two ends of the admittance semicircle, which means that the diameter of the semicircle is equal to $1/R$. But why an *upside-down* semicircle? Time for our next Theory Note.

Theory Note 2.3. Why the Admittance Semicircle Is 'Upside Down'

The impedance of the series RC circuit was given by equation (2.8). To get the admittance, we need to find $1/Z$.

$$Y_{\text{total}} = \frac{1}{Z_{\text{total}}} = \frac{1}{R - \frac{j}{\omega C}} = \frac{\omega C}{R\omega C - j} \tag{2.11}$$

$$= \frac{\omega C(R\omega C + j)}{1 + R^2\omega^2 C^2} = \frac{R\omega^2 C^2}{1 + R^2\omega^2 C^2} + j\frac{\omega C}{1 + R^2\omega^2 C^2}$$

Here, we used our trick with the *complex conjugate* again. By multiplying top and bottom by $(R\omega C + j)$, we obtain the real and imaginary parts of the admittance.

$$\text{Re}\,[Y_{\text{total}}] = \frac{R\omega^2 C^2}{1 + R^2\omega^2 C^2} \quad \text{Im}\,[Y_{\text{total}}] = \frac{\omega C}{1 + R^2\omega^2 C^2} \tag{2.12}$$

Note that the imaginary component of the admittance is positive. At some frequency, the real and imaginary parts of the admittance will be the same. This means that the phase angle will be $+45°$ or $+\pi/4$ (the admittance plot is in the $+j$ quadrant of the complex plane, whereas the impedance plot is in the $-j$ quadrant).

$$\frac{R\omega^2 C^2}{1 + R^2\omega^2 C^2} = \frac{\omega C}{1 + R^2\omega^2 C^2} \quad \omega = \frac{1}{RC} \tag{2.13}$$

Since the real and imaginary components of the admittance are equal at the radial frequency $\omega_0 = 1/RC$, this must correspond to the bottom of the admittance semicircle in Figure 2.16.

So, the impedance and admittance both have a 45° phase shift for $\omega = 1/RC$. The *sign* of the phase angle is different (negative for Z and positive for Y), and the *shape* of the plots is different as shown in Figure 2.17.

It is easy to show that impedance and admittance, when expressed in polar coordinates, are related by $|Y| = 1/|Z|$ and $\phi_Y = -\phi_Z$. I have left this for you to solve as one of the problems at the end of the chapter.

(Continued)

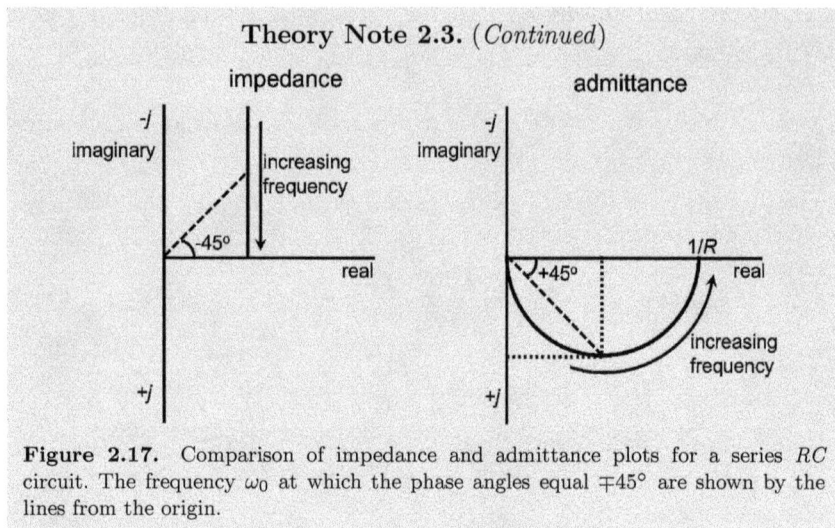

Figure 2.17. Comparison of impedance and admittance plots for a series RC circuit. The frequency ω_0 at which the phase angles equal $\mp 45°$ are shown by the lines from the origin.

2.6 Parallel Connections

Now that we have looked at series connections, we can turn to parallel connections, starting with the parallel RC connection shown in Figure 2.18.

The resistor and capacitor values are the same as those we used for the series RC connection. When we build the circuit in ZView® and run the simulator, we get the result shown in Figure 2.19.

Now the impedance response looks like the admittance response for the series RC circuit. Can we rationalize this with our 'two simple questions' approach? The answer is yes. For zero frequency (dc), the capacitor blocks current flow, so the current can only flow through the resistor. The zero-frequency impedance must therefore be equal to R ($10^3\Omega$), and it must lie on the real axis. What about the infinite frequency limit? The impedance of the capacitor is then zero, so that it short circuits out the resistor. The high-frequency intercept is therefore at the origin. So, we

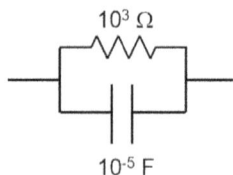

Figure 2.18. A parallel RC connection.

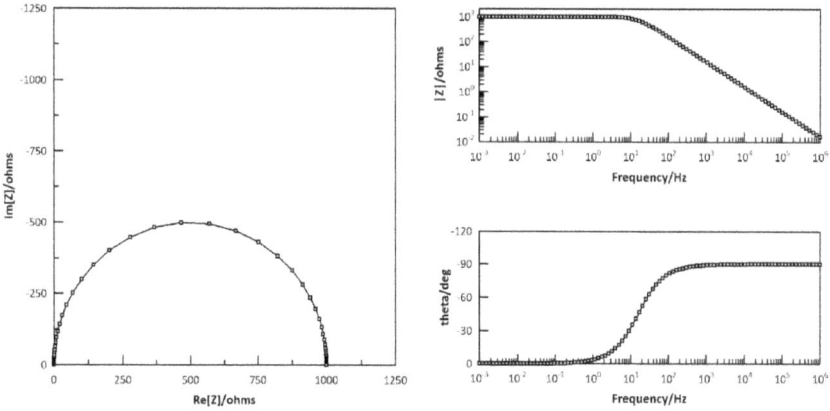

Figure 2.19. Impedance response of the parallel RC circuit shown in Figure 2.18.

have two points, but again why a semicircle in between? Time for another Theory Note.

Theory Note 2.4. Impedance and Admittance of a Parallel RC Circuit

We have a parallel circuit. So, we need to *add the inverses* of the impedances Z_R and Z_C (i.e., add the admittances).

$$\frac{1}{Z_{\text{total}}} = \frac{1}{Z_R} + \frac{1}{Z_C} = \frac{1}{R} + j\omega C \tag{2.14}$$

To find Z_{total} we need to invert $1/Z_{\text{total}}$. Let us see what happens when we do this.

$$Z_{\text{total}} = \frac{1}{\frac{1}{R} + j\omega C} = \frac{R}{1 + j\omega RC} = \frac{R(1 - j\omega RC)}{1 + \omega^2 R^2 C^2} \tag{2.15}$$

Note that we used the *complex conjugate* again to obtain the real and imaginary parts of the total impedance. They are

$$\text{Re}[Z_{\text{total}}] = \frac{R}{1 + \omega^2 R^2 C^2} \qquad \text{Im}[Z_{\text{total}}] = -\frac{\omega R^2 C}{1 + \omega^2 R^2 C^2} \tag{2.16}$$

Now the magnitudes of Z_R and Z_C will be equal at a particular frequency, i.e., when $R = 1/\omega_0 C$ or $\omega_0 = 1/RC$. Half of the current will go through the resistor and half, through the capacitor. When the

(Continued)

Theory Note 2.4. (*Continued*)

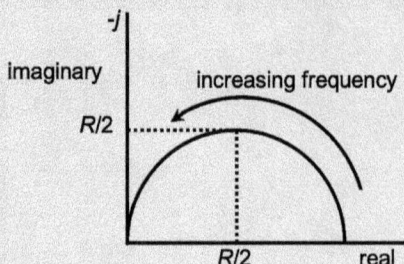

Figure 2.20.   Semicircular Nyquist impedance plot for a parallel RC network.

radial frequency $\omega_0 = 1/RC$, the real and imaginary parts of Z_{total} both become $R/2$ as shown in Figure 2.20.

$$\text{Re}[Z_{\text{total}}] = \frac{R}{1+1} = +\frac{R}{2} \quad \text{Im}\,[Z_{\text{total}}] = -\frac{R}{1+1} = -\frac{R}{2} \quad (2.17)$$

Now we will take a closer look at the *admittance* of the parallel circuit. We start by asking our two common-sense questions. What will the admittance be at dc (zero frequency)? Easy – the admittance of the capacitor is zero (it blocks dc current). So, we are just left with the admittance of the resistor, which is $1/R$. What about the admittance at infinite frequency? Since the impedance of the capacitor becomes zero at infinite frequency, its admittance must be infinite. So, our total admittance will tend towards infinity as the frequency increases. No semicircle here then. Let us see what ZView® tells us (Figure 2.21).

You can see that the Nyquist plot of admittance for a parallel RC circuit looks a lot like the Nyquist impedance plot for a series RC circuit.

2.7 Tackling More Complicated Networks

By now you should be getting a good feel for the approach that we are taking. Asking our common-sense questions and then getting ZView® to calculate the plots. So, we can be a bit more ambitious and tackle a three-element network (see Figure 2.22).

We start with our usual two questions. At zero frequency (dc), the current must flow through R_1 and R_2 because the capacitor C blocks dc. At infinite frequency, capacitor C shorts out resistor R_2, so only resistor R_1 is left. What about the special case where half the current flows through

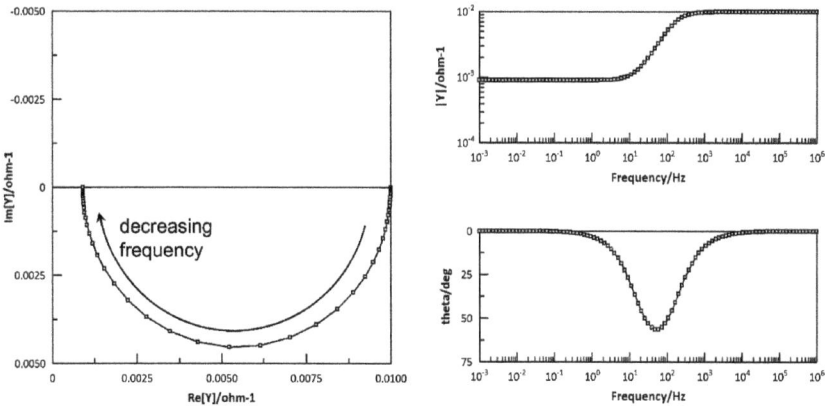

Figure 2.21. Admittance plots for the parallel RC circuit ($R = 10^3\,\Omega, C = 10^{-5}\,\mathrm{F}$). Note that the zero-frequency intercept on the real axis corresponds to $1/R$, in this case, $1/10^3\,\Omega = 10^{-3}\,\Omega^{-1}$.

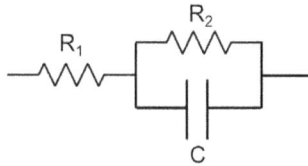

Figure 2.22. A series/parallel RC circuit.

R_2 and half, through C? We have encountered this before for the parallel circuit. It occurs when the radial frequency $\omega_0 = 1/R_2 C$ and the real and imaginary components of the parallel part of the circuit are equal. It is worth working through this example, so here is another Theory Note.

Theory Note 2.5. Impedance of a Series/Parallel RC Circuit (Figure 2.22)

We have already worked out the impedance of the parallel part of the circuit. Look back at the last Theory Note, where we showed that

$$\mathrm{Re}\,[Z_{\mathrm{parallel}}] = \frac{R_2}{1 + \omega^2 R_2^2 C^2} \qquad \mathrm{Im}\,[Z_{\mathrm{parallel}}] = -\frac{\omega R_2^2 C}{1 + \omega^2 R_2^2 C^2} \qquad (2.18)$$

(*Continued*)

Theory Note 2.5. (*Continued*)

All we need to do now is add the impedance of the series resistor R_1 to obtain

$$\text{Re}\,[Z_{\text{total}}] = R_1 + \frac{R_2}{1 + \omega^2 R_2^2 C^2} \qquad \text{Im}[Z_{\text{total}}] = -\frac{\omega R_2^2 C}{1 + \omega^2 R_2^2 C^2} \quad (2.19)$$

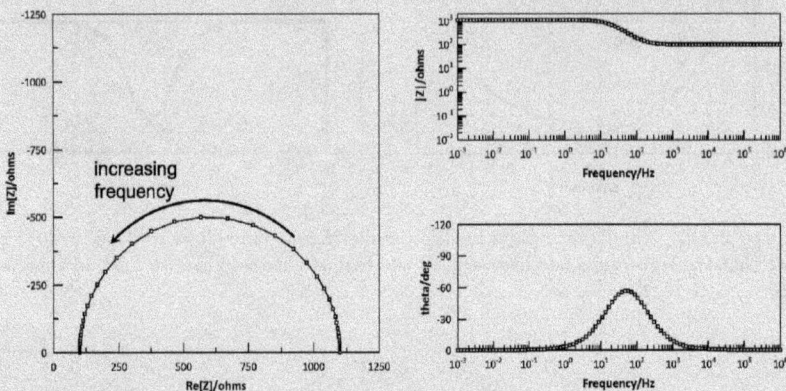

Figure 2.23. Impedance plots for the series/parallel circuit in Figure 2.22. $R_1 = 100\,\Omega, R_2 = 10^3\,\Omega, C = 10^{-5}$ F.

So, all the series resistor does is displace the real component along the real axis by R_1. ZView® confirms this – see Figure 2.23.

We see that the high-frequency intercept is indeed at $R_1(100\,\Omega)$, and the zero-frequency intercept is at $R_1 + R_2(1100\,\Omega)$. The *diameter of the semicircle* is equal to the parallel resistance R_2. The phase angle is interesting. Instead of changing from 0 to $-90°$ as in the case of the parallel *RC* circuit with no series resistor, here *the phase angle shows a peak*. This is because the high-frequency limit is R_1 rather than the origin, where the impedance becomes real (zero phase angle). It is important to note that *the maximum phase angle does not occur at the radial frequency $\omega = 1/R_2C$*. It is a higher frequency as can be seen by drawing a tangent to the semicircle from the origin as shown in Figure 2.24. You can see that the more we displace the semicircle along the real axis by increasing the series resistance R_1, the lower the maximum phase angle will be.

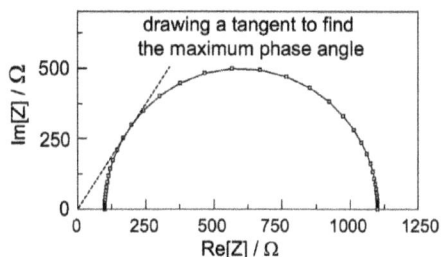

Figure 2.24. Drawing a tangent to find the frequency at which the phase angle is maximum.

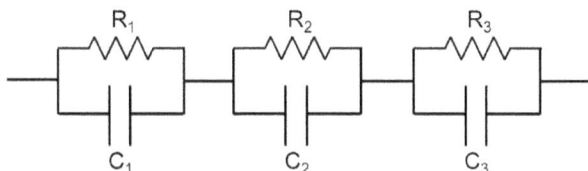

Figure 2.25. Circuit with three parallel RC sub-circuits with time constants R_1C_1, R_2C_2, R_3C_3 that differ by a factor of 10. $R_1 = R_2 = R_3 = 10^3\,\Omega, C_1 = 10^{-7}\,\text{F}, C_2 = 10^{-6}\,\text{F}, C_3 = 10^{-5}\,\text{F}$. The corresponding RC time constants are $10^{-4}\,\text{s}$, $10^{-3}\,\text{s}$ and $10^{-2}\,\text{s}$.

We can use the same approach to look at more complex RC circuits like the one in Figure 2.25, which has three parallel RC sub-circuits connected one after the other.

Here, we have chosen values of the capacitors to ensure that the time constants of the three parallel sub-circuits differ by factors of 10. As Figure 2.26 shows, in this case, three semicircles corresponding to the three sub-circuits can be discerned in the Nyquist plot, although the overlap is substantial. Starting from the lowest frequency semicircle on the right, the plot reflects the order of the RC time constants $10^{-2}\,\text{s}$, $10^{-3}\,\text{s}$ and $10^{-4}\,\text{s}$.

Now that we have the general idea of how to handle RCL networks, we can have some fun. Can we predict the impedance of the circuit shown in Figure 2.27?

This looks interesting. We have an inductor and a capacitor in series in the bottom branch of the circuit. The impedance of the inductor goes up with frequency and the impedance of the capacitance goes down. So, at low frequencies, the circuit will look like a parallel RC network, because the

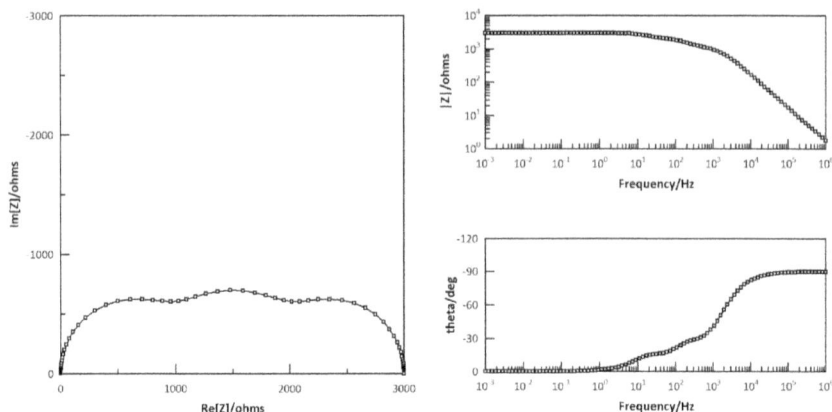

Figure 2.26. Impedance plots for the equivalent circuit in Figure 2.25. Note that the three semicircles overlap substantially, even though the three time-constants differ by factors of 10.

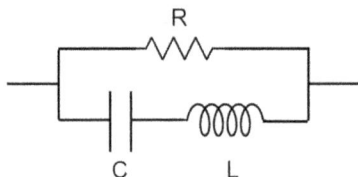

Figure 2.27. A brain teaser. What will the Nyquist and Bode plots look like for this circuit?

impedance of the inductor will be small. At high frequencies, the inductor will shut down the bottom branch and the current will be forced to flow through the resistor. So, let us cut the suspense and see what ZView® gives us. Figure 2.28 shows the values used in the simulation and Figure 2.29 shows the result.

Interesting! The Nyquist plot is a circle. It starts at 1000 Ω at low frequencies and then presents a semicircle in the upper complex plane. This is the behaviour we expect for a parallel RC circuit. However, instead of stopping at the origin, as we would expect if there were no inductor, *the plot carries on through to the lower quadrant*. This is because the impedance of the inductor is now increasing sufficiently to begin to 'choke off' the current through the lower branch of the circuit. We then go around the parallel

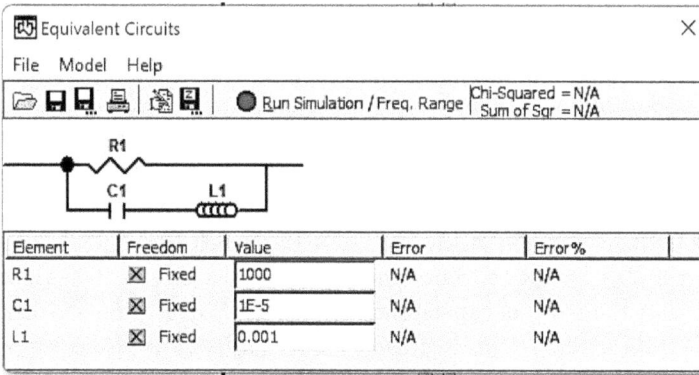

Figure 2.28. Values input into ZView® for the circuit in Figure 2.27.

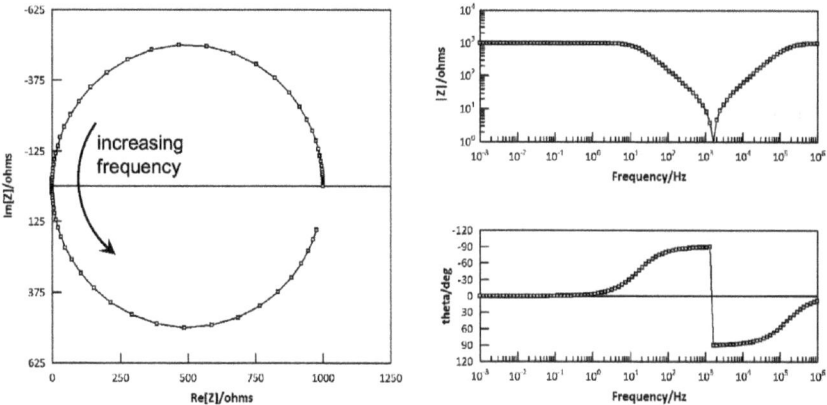

Figure 2.29. The impedance response of the *RLC* circuit in Figure 2.27.

RL semicircle back to 1000 Ω. Now all the current is flowing through *R*, just as it was at dc. This circuit has the interesting property of allowing current through very easily for a particular frequency. It is an example of an electronic device known as a *tuned filter* of the type that were used in radio sets receiving amplitude-modulated signals at a fixed frequency – the old 'long wave' programs.

2.8   A Health Warning: Your Circuit May Not Be Unique!

By now, you should be getting good at guessing which equivalent circuit to use to describe your experimental results. Have a look at the simulated impedance response shown in Figure 2.30. This looks straightforward. The high-frequency intercept suggests a series resistance of 100 Ω, and the semicircle looks like a typical parallel RC response with a resistance of 1000 Ω. The frequency of the maximum is 166 Hz, which corresponds to a time constant $1/2\pi f = 10^{-3}$ s. Since the parallel resistance is 1000 Ω, we can deduce that the parallel capacitance is 10^{-6} F. Now for the shocker. *This response was actually simulated using a quite different equivalent circuit.* It is shown on the right-hand side of Figure 2.31.

The problem of circuit degeneracy is more general. Stephen Fletcher (Fletcher, 1994) has summarized degenerate circuits containing three, four, five or six elements and given expressions to relate the values of the elements. He has also shown that degeneracy applies to distributed impedances as well. This paper gives algebraic expressions that allow conversion of one circuit with specified impedance values into the other degenerate circuit(s) with a different set of values. The paper is well worth reading to alert you to the dangers of degenerate circuits. An explanation of how the degeneracy condition gives rise to the values of

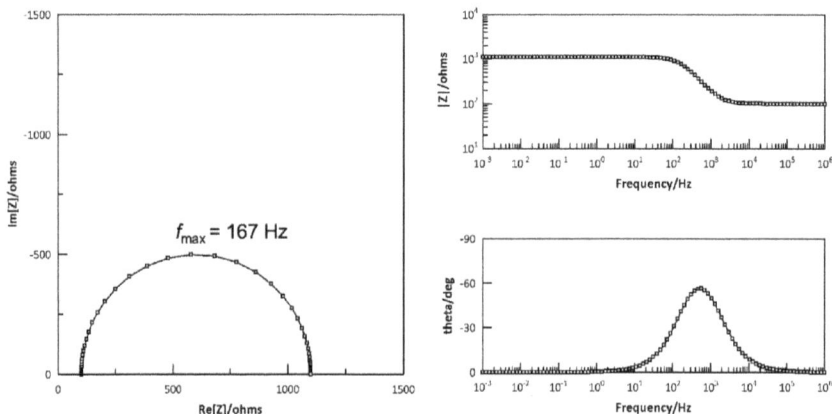

Figure 2.30.   A simulated impedance response that looks straightforward but is not. Although you will probably think that the equivalent circuit is the one shown on the left-hand side of Figure 2.31, in fact *the plots were simulated using the circuit on the right-hand side.* The two equivalent circuits are said to be *degenerate* because for given values they are indistinguishable.

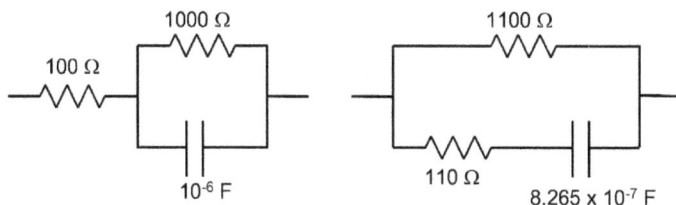

Figure 2.31. The R and C values for two degenerate equivalent circuits. The two different circuits give an *identical* frequency response.

resistance and capacitance for the right-hand circuit in Figure 2.30 is given in Appendix 2.2.

So, how do we deal with the problem of degeneracy when analyzing experiments? The only way forward in this situation is to assign physical significance to the elements in the possible circuits and then to look at how the values of the elements vary when the experimental conditions are altered. If the values change in a systematic way that is consistent with theory (e.g., the dependence of the Faradaic resistance on concentration and potential), then we can be reasonably sure that we have the right circuit. In the same way if, for example, the values of an element change in an unexpected way, for example, the double layer capacitance changes with the concentration of redox species, we can suspect that we have the wrong circuit. A key point here is that *restricting impedance measurements to one set of experimental conditions can lead to uncertainty in the interpretation because of circuit degeneracy.*

Now it is time to leave the discussion of RCL circuits and in the next chapter we begin to look in more detail at the impedance responses corresponding to electrochemical processes.

Problems

2.1. Referring to Figure 2.10 (and without looking at Appendix 2.1!), follow the stepwise procedure to derive an expression for the total impedance of the circuit. Did you get the same result as in Appendix 2.1?

2.2. Show that in general the impedance and admittance of any RCL circuit are related in polar coordinates (magnitude, phase angle) by

$$|Y| = \frac{1}{|Z|} \quad \phi_Y = -\phi_Z$$

2.3. Look at the following Nyquist and Bode plots below and decide what kind of equivalent circuit is appropriate to fit the data. Estimate the values of the components in your circuit. Pay attention to the axis labels!

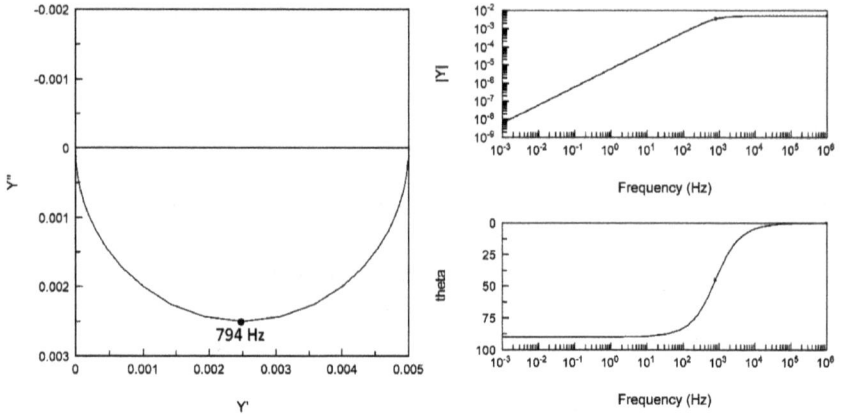

2.4. Look at the following Nyquist and Bode plots and decide what kind of equivalent circuit is appropriate to fit the data. Estimate the values of the components in your circuit.

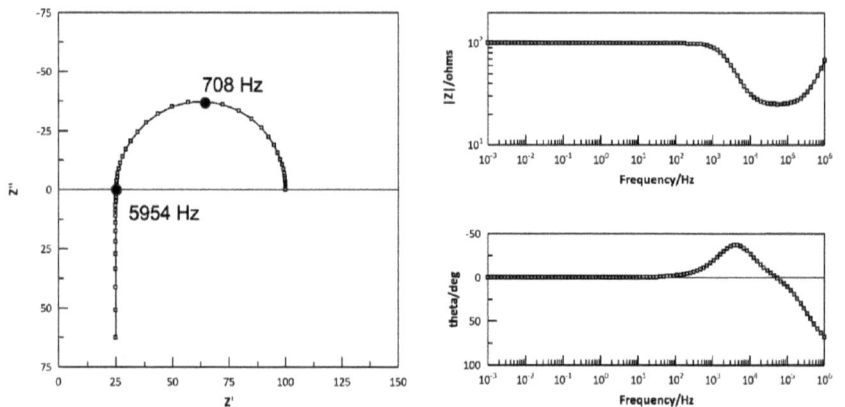

2.5. Look at the following Nyquist and Bode plots and decide what kind of equivalent circuit is appropriate to fit the data. Estimate the values of the components in your circuit.

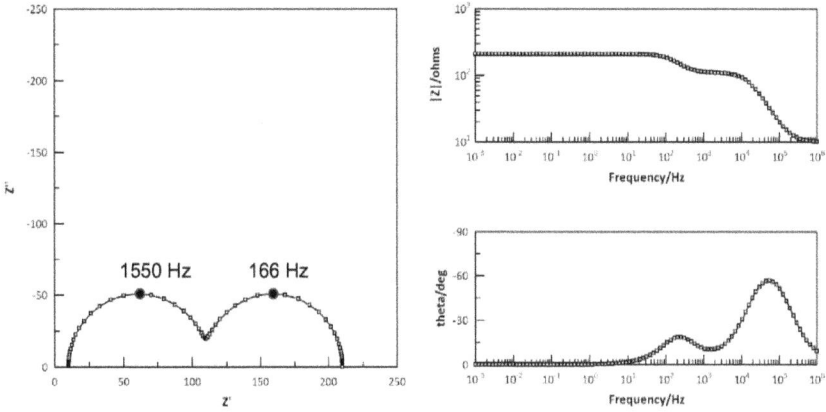

Appendix 2.1 Working Out the Impedance of the Circuit in Figure 2.10 (Shown Again in Figure A2.1)

$$Z_A = R_3 - \frac{j}{\omega C_2} = \frac{\omega R_3 C_2 - j}{\omega C_2} \tag{A2.1}$$

$$\frac{1}{Z_A} = \frac{\omega C_2}{\omega R_3 C_2 - j} = \frac{\omega C_2 \left(\omega R_3 C_2 + j\right)}{1 + \omega^2 R_3^2 C_2^2}$$

$$= \frac{\omega^2 R_3 c_2^2 + j\omega C_2}{1 + \omega^2 R_3^2 C_2^2} \tag{A2.2}$$

$$\mathrm{Re}\left(\frac{1}{Z_A}\right) = \frac{\omega^2 R_3 c_2^2}{1 + \omega^2 R_3^2 C_2^2} \qquad \mathrm{Im}\left(\frac{1}{Z_A}\right) = \frac{\omega C_2}{1 + \omega^2 R_3^2 C_2^2} \tag{A2.3}$$

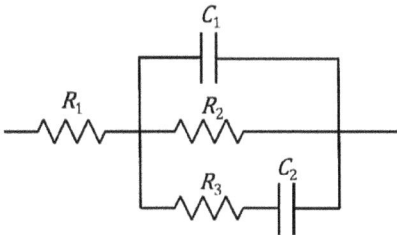

Figure A2.1. *RC* circuit in Figure 2.10.

$$\frac{1}{Z_B} = \frac{1}{Z_A} + \frac{1}{R_2} + j\omega C_1$$

$$= \mathrm{Re}\left(\frac{1}{Z_A}\right) + \frac{1}{R_2} + j\,\mathrm{Im}\left(\frac{1}{Z_A}\right) + j\omega C_1 \qquad (A2.4)$$

$$\mathrm{Re}\left(\frac{1}{Z_B}\right) = \mathrm{Re}\left(\frac{1}{Z_A}\right) + \frac{1}{R_2} \quad \mathrm{Im}\left(\frac{1}{Z_B}\right) = \mathrm{Im}\left(\frac{1}{Z_A}\right) + \omega C_1 \quad (A2.5)$$

$$\mathrm{Re}\left(\frac{1}{Z_B}\right) = \frac{\omega^2 R_3 c_2^2}{1 + \omega^2 R_3^2 C_2^2} + \frac{1}{R_2} \quad \mathrm{Im}\left(\frac{1}{Z_B}\right)$$

$$= \frac{\omega C_2}{1 + \omega^2 R_3^2 C_2^2} + \omega C_1 \qquad (A2.6)$$

$$Z_B = \frac{1}{\mathrm{Re}\left(\frac{1}{Z_B}\right) + j\,\mathrm{Im}\left(\frac{1}{Z_B}\right)}$$

$$= \frac{\mathrm{Re}\left(\frac{1}{Z_B}\right) - j\,\mathrm{Im}\left(\frac{1}{Z_B}\right)}{\mathrm{Re}^2\left(\frac{1}{Z_B}\right) + \mathrm{Im}^2\left(\frac{1}{Z_B}\right)} \qquad (A2.7)$$

$$\mathrm{Re}(Z_B) = \frac{\mathrm{Re}\left(\frac{1}{Z_B}\right)}{\mathrm{Re}^2\left(\frac{1}{Z_B}\right) + \mathrm{Im}^2\left(\frac{1}{Z_B}\right)}$$

$$\mathrm{Im}\,(Z_B) = -\frac{\mathrm{Im}\left(\frac{1}{Z_B}\right)}{\mathrm{Re}^2\left(\frac{1}{Z_B}\right) + \mathrm{Im}^2\left(\frac{1}{Z_B}\right)} \qquad (A2.8)$$

$$\mathrm{Re}\,(Z_\mathrm{total}) = R_1 + \mathrm{Re}\,(Z_B) = R_1 + \frac{\mathrm{Re}\left(\frac{1}{Z_B}\right)}{\mathrm{Re}^2\left(\frac{1}{Z_B}\right) + \mathrm{Im}^2\left(\frac{1}{Z_B}\right)} \qquad (A2.9)$$

$$\mathrm{Im}\,(Z_\mathrm{total}) = \mathrm{Im}\,(Z_B) = -\frac{\mathrm{Im}\left(\frac{1}{Z_B}\right)}{\mathrm{Re}^2\left(\frac{1}{Z_B}\right) + \mathrm{Im}^2\left(\frac{1}{Z_B}\right)} \qquad (A2.10)$$

$$\mathrm{Re}\,(Z_\mathrm{total}) = R_1 + \frac{\frac{\omega^2 R_3 c_2^2}{1+\omega^2 R_3^2 C_2^2} + \frac{1}{R_2}}{\left(\frac{\omega^2 R_3 c_2^2}{1+\omega^2 R_3^2 C_2^2} + \frac{1}{R_2}\right)^2 + \left(\frac{\omega C_2}{1+\omega^2 R_3^2 C_2^2} + \omega C_1\right)^2}$$

$$(A2.11)$$

$$\text{Im}\left(Z_{\text{total}}\right) = \frac{\frac{\omega C_2}{1+\omega^2 R_3^2 C_2^2} + \omega C_1}{\left(\frac{\omega^2 R_3 c_2^2}{1+\omega^2 R_3^2 C_2^2} + \frac{1}{R_2}\right)^2 + \left(\frac{\omega C_2}{1+\omega^2 R_3^2 C_2^2} + \omega C_1\right)^2} \qquad \text{(A2.12)}$$

Here is a spreadsheet transform that I used to calculate the impedance.

'Input values
R1 = 100
R2 = 1000
R3 = 10000
C1 = 1e-6
C2 = 1e-5
pi = 3.14159

'set up frequency range from 1 mHz to 1 MHz with ten steps per decade

logf = data(−3,6,0.1)
col(1) = 10^logf
f = col(1)
omega = 2*pi*f

'Step 1
ReYA = omega^2*R3*C2^2/(1 + omega^2*R3^2*C2^2)
ImYA = omega*C2/(1 + omega^2*R3^2*C2^2)
col(2) = ReYA
col(3) = ImYA

'Step 2
ReYB = ReYA + (1/R2)
ImYB = ImYA + (omega*C1)
col(4) = ReYB
col(5) = ImYB

'Step 3
ReZB = ReYB/(ReYB^2 + ImYB^2)
ImZB = −ImYB/(ReYB^2 + ImYB^2)

'plot in complex plane
col(6) = ReZB
col(7) = ImZB
 The result of the calculation is shown in Figure A2.2.

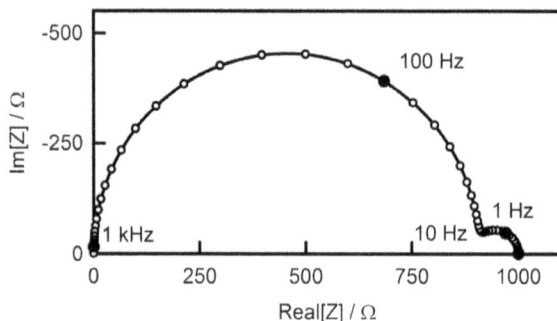

Figure A2.2. Impedance calculated for the circuit 2.10 with the following values: $R_1 = 100\,\Omega$, $R_2 = 10^3\,\Omega$, $R_3 = 10^4\,\Omega$, $C_1 = 10^{-6}$ F, $C_2 = 10^{-5}$ F. See what follows for the explanation of the two semicircles and the frequencies of the two maxima.

We can apply our common-sense approach to try to understand where the two semicircles come from. At high frequencies, the impedance of the capacitor C_2 becomes small, so we are left with $R_2(10^3\,\Omega)$, $R_3(10^4\,\Omega)$ and $C_1(10^{-6}$ F) in parallel. $10^3\,\Omega$ and $10^4\,\Omega$ in parallel correspond to $R_{\text{tot}} = 909\,\Omega$. The high-frequency RC time constant is therefore $909\,\Omega \times 10^{-6}\text{F} = 9.09 \times 10^{-4}$ s. Since $\omega = 2\pi f = 1/RC = 1.1 \times 10^3\,\text{s}^{-1}$, we find $f_{\text{HF}} = 175$ Hz. At low frequencies, the capacitors have high impedances, so we can forget R_3 and are left with a parallel combination of C_1, R_2 and C_2. Since the capacitances add in parallel, $C_{\text{tot}} = 1.1 \times 10^{-5}$ F. The low-frequency RC time constant is therefore $1.1 \times 10^{-5}\,\text{F} \times 10^3\,\Omega = 1.1 \times 10^{-2}$ s, corresponding to $\omega_{\text{LF}} = 90.9\,\text{s}^{-1}$. The low-frequency maximum therefore occurs at $f_{\text{LF}} = 14.5$ Hz.

Appendix 2.2   Transforming the Values of the Elements in Degenerate Circuits (See Figure A2.3)

For circuit A, we have $R_{A,1} = 100\,\Omega$, $R_{A,2} = 1000\,\Omega$ and $C_A = 10^{-6}$ F.

We begin by defining the following impedance ratios for the two circuits.

$$\frac{R_{A,1}}{R_{A,2}} = a \quad \frac{R_{B,2}}{R_{A,2}} = b \quad \frac{C_A}{C_B} = c \quad \frac{R_{B,1}}{R_{A,2}} = d \qquad (\text{A2.13})$$

For degenerate circuits of this type, Fletcher (Fletcher, 1994) gives the following equalities:

$$b = a + a^2 \quad c = (1 + a)^2 \quad d = 1 + a \qquad (\text{A2.14})$$

Using these equalities, we can convert circuit A to circuit B.

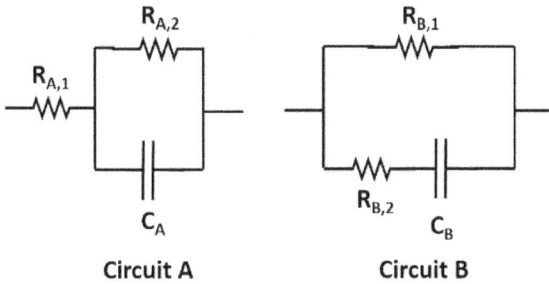

Figure A2.3. Two degenerate circuits.

From $b = a + a^2$, we obtain

$$\frac{R_{B,2}}{R_{A,2}} = \frac{R_{A,1}}{R_{A,2}} + \left(\frac{R_{A,1}}{R_{A,2}}\right)^2 \text{ so that } R_{B,2} = R_{A,1} + \frac{R_{A,1}^2}{R_{A,2}} \qquad (A2.15)$$

So that $R_{B,2} = 100 + (100)^2/1000 = \underline{110\,\Omega}$.

From $c = (1 + a)^2$, it follows that $\dfrac{C_A}{C_B} = \left(1 + \dfrac{R_{A,1}}{R_{A,2}}\right)^2$ \qquad (A2.16)

So that $C_B = 10^{-6}/(1 + 0.1)^2 = 10^{-6}/1.21 = \underline{8.26 \times 10^{-7}\text{F}}$.
Finally, from $d = 1 + a$, $\frac{R_{B,1}}{R_{A,2}} = 1 + \frac{R_{A,1}}{R_{A,2}}$ so that

$$R_{B,1} = R_{A,2} + R_{A,1} \qquad (A2.17)$$

Substituting the values: $R_{B,1} = 1000 + 100 = \underline{1100\,\Omega}$.

We have obtained values of the three elements in circuit B that make it degenerate with circuit A.

Reference

Fletcher, S. 1994. Tables of degenerate electrical networks for use in the equivalent-circuit analysis of electrochemical systems. *Journal of the Electrochemical Society*, 141, 1823–1826.

Chapter 3

Putting the E in EIS: Frequency-Response Analysis of Electrochemical Systems

3.1 Introduction

In this chapter, we begin to look more closely at how electrochemical systems and devices can be modelled using what we have learned about *RCL* circuits. We will see that some new types of circuit elements are required to describe diffusion and porous electrodes. Of course, this chapter is not intended to be an authoritative text on electrochemistry. There are many excellent textbooks that provide the information in much more depth (see *Further Reading* at the end of this chapter). Nevertheless, some of the key information is summarized in this chapter, because it is needed to understand the impedance behaviour of electrochemical systems.

3.2 The Electrical Double Layer

The contact between an electronic conductor like a metal and an ionic conductor (electrolyte solution or solid ion conductor) is said to be *ideally polarizable* if electrons and ions are unable to cross the interface. This means that no dc current flows. In practice, the behaviour of the interface will depend on the nature of the electrode, the applied potential, and the composition of the ionic phase. In the case of oxygen-free aqueous solutions of common salts like NaCl, there is a quite large *potential window* of 1.23 V between the potentials for water reduction and water oxidation. Provided that the electrode (usually a metal) cannot be oxidized at potentials inside this window, it can approximate very well to an ideally polarizable electrode. In the case of mercury electrodes, the window is even wider because the hydrogen evolution reaction is extremely slow.

The excess electronic charge on a metal electrode immersed in an inert electrolyte solution is balanced by an equal and opposite net ionic charge in the electrolyte associated with an excess of ions of one type (cation or anion). This arrangement of charge is called the *electrical double layer*. For concentrated electrolytes, the *Helmholtz model* of the double layer pictures the excess ionic charge as forming a layer of counter charge separated from the metal by a layer of solvent molecules as illustrated in Figure 3.1 (this is a simplified view of the electrical double layer that ignores the diffuse double layer and factors such as adsorption of ions, but it should suffice for the discussion of EIS).

In the case of a normal capacitor, the charge is a linear function of voltage, but this is not the case for the double-layer capacitance, which varies with electrode potential. However, for any potential, we can define the (potential-dependent) *differential double-layer capacitance* as

$$C_{dl} = \left(\frac{dQ_m}{dE} \right) \tag{3.1}$$

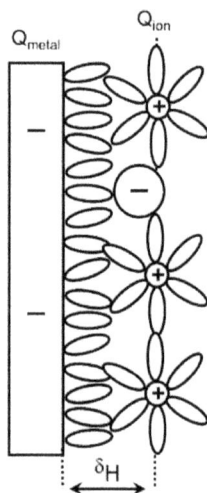

Figure 3.1. The Helmholtz double layer illustrated for the situation where a negative electronic charge on the electrode is balanced by an equal and opposite excess positive ionic charge situated at a distance δ_H. The simple parallel plate capacitor model predicts a double-layer capacitance of around $30\,\mu F\ cm^2$.

where Q_m is the charge on the metal and E is the electrode potential. Since we use small amplitudes in our impedance measurements, the measured capacitance corresponds to the differential double-layer capacitance, but we will drop the 'differential' bit from now on.

To get an idea of the magnitude of C_{dl}, suppose that the 'plates' of our Helmholtz capacitor are separated by a distance δ_H equivalent to the dimensions of two water molecules (around 0.3 nm). Because the plates are so close together, the electric field between them is high – of the order of 10^7 V cm^{-1}. Under these conditions, the dipoles of the water molecules are almost fully oriented and the rotational contribution to the dielectric constant is 'frozen out'. This means that the relative permittivity ε_{H_2O} is between 10 and 20, rather than the bulk value of 80 at room temperature. Using $\varepsilon_{H_2O} = 10$ in the capacitance equation, $C_{dl} = \dfrac{\varepsilon_{H_2O}\varepsilon_0}{\delta_H} \approx 30\,\mu\text{F cm}^{-2}$.

3.3 The Charge Transfer Resistance

Now we suppose we add electroactive species to our electrolyte. The interface is no longer ideally polarizable. We can still charge the double layer, but electrons can now leak across the interface in both directions. In other words, an *electrode reaction* occurs. If equal concentrations of oxidized and reduced species are present in the solution, then no current will flow at the *equilibrium potential* given by the Nernst equation (named after Walter Nernst, who received the 1920 Nobel Prize for the formulation of the Third Law of Thermodynamics: reportedly his office and laboratory were so untidy that his co-workers defined them as being in the state of maximum entropy). The Nernst equation is normally written in terms of *standard reduction potential*, E^o, and the thermodynamic *activities* a_O and a_R of the oxidized and reduced species

$$E_{eq} = E^o + \frac{RT}{nF}\ln\frac{a_O}{a_R} \tag{3.2a}$$

For the purposes of this book, we will use the alternative form where activities are replaced by concentrations, and the standard reduction potential is replaced by the *formal potential* $E^{o'}$.

$$E_{eq} = E^{o'} + \frac{RT}{nF}\ln\frac{C_O}{C_R} \tag{3.2b}$$

Here, C_O and C_R are the concentrations of the oxidized and reduced species, O and R. If the electrode potential is more positive than the equilibrium potential, oxidation will occur. If the potential is more negative than E_{eq}, reduction will take place. If diffusion is so fast that the rate of the reaction is controlled only by the kinetics of electron transfer, the current density j (not to be confused with j, the square root of -1) is given by the *Butler–Volmer equation*. The equation is also called the Erdey-Grúz–Volmer equation. The basic idea originated with the UK physical chemist John Alfred Valentine Butler in the 1930s, but at around the same time it was expressed more clearly by Max Volmer and Tibor Erdey-Grúz.

$$j = j_0 \left[\exp\left(\frac{(1-\alpha)nF(E - E_{eq})}{RT} \right) - \exp\left(-\frac{\alpha nF(E - E_{eq})}{RT} \right) \right] \quad (3.3)$$

where α is the cathodic transfer coefficient, and j_0, the exchange current density, is defined as

$$j_0 = nFk^\circ C_O^\alpha C_R^{1-\alpha} \quad (3.4)$$

Here, k° is the *standard heterogeneous rate constant* for the reaction.

Returning to equation (3.3), the exponential terms represent the rates of the oxidation and reduction reactions, respectively. You can see that when the *overpotential* $E - E_{eq}$ is zero, both exponential terms cancel out ($e^0 = 1$), so the current is zero. The overpotential is given the symbol η in most texts. Figure 3.2 shows the typical shape of the Butler–Volmer current potential plot.

We will suppose that we have equal concentrations of O and R and set the dc potential to be equal to the equilibrium potential. No dc current flows. Now we perturb the system with a small amplitude sinusoidal voltage symmetrically about E_{eq}. What will the current response be if the kinetics

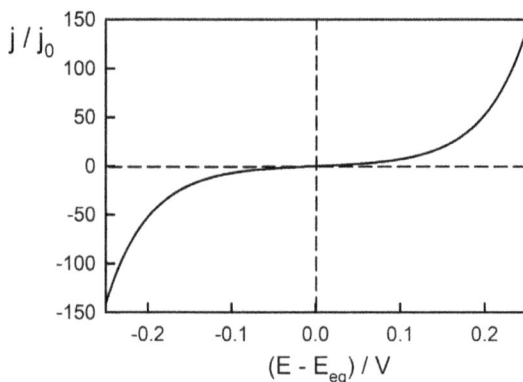

Figure 3.2. Plot of the current density j vs. overpotential, $E - E_{eq}$ predicted by the Butler–Volmer equation ($n = 1, \alpha = 0.5$).

are determined only by the Butler–Volmer equation, i.e., if we can neglect the effects of diffusion? Time for our first Theory Note of this chapter.

Theory Note 3.1. Linearizing the Butler–Volmer Equation

'Small perturbation' means that the overpotential $E - E_{eq}$ in the exponential terms is so small that we can use the approximations $e^x = 1 + x$ and $e^{-x} = 1 - x$. In practice, this means that we use a perturbation of only a few mV. Then substituting these approximations in the Butler–Volmer equation – equation (3.3) – we find that

$$j = j_0 \frac{nF\eta}{RT} \qquad (3.5)$$

This means that the current density depends linearly on the overpotential. What we have done here is equivalent to drawing a tangent to the Butler–Volmer plot at the equilibrium potential.

We can rearrange equation (3.5) to obtain the *charge transfer resistance* or *Faradaic resistance* as

$$R_F = \frac{d\eta}{dj} = \frac{RT}{nFj_0} \qquad (3.6)$$

R_F is a linear circuit element that we can use when modelling the EIS response of electrode processes.

3.4 Mass Transport in Electrochemistry

So far, we have only discussed what happens when electrons are transferred from or to reduced/oxidized (redox) species located at the electrode surface. The calculation of the current–voltage plot from the Butler–Volmer equation – see Figure 3.2 – assumes that the concentrations of oxidized and reduced species are both constant. However, this will not be the case generally because the flow of current will alter the concentrations of O and R close to the surface as the result of consumption of the reactants and generation of products. The *mass transport* of reactants and products is therefore important in determining the local concentration of species at the electrode surface. Three modes of mass transport need to be considered: diffusion, migration and convection. Diffusion is the consequence of random thermal motion of molecules that leads to the transfer of material from regions of high concentration to regions of lower concentration. Migration involves the movement of charged ions in an electric field. Finally, convection involves movement of the liquid as a whole, resulting in the transport of dissolved species. The effects of migration in the study of electrode reactions involving solution species can be minimized by using a high concentration of an unreactive salt – the 'supporting electrolyte' – and a much lower concentration of electroactive species. In this case, the dominant mode of transport of the material of interest is diffusion. However, in many practical applications of electrochemistry (batteries, electrolyzers, fuel cells), migration is important and needs to be accounted for in the analysis of experimental data. Convection is quite different since it involves the flow of liquid. Natural convection occurs because of gradients of density or temperature. Forced convection involves some mechanical input such as stirring or flowing the liquid. The *rotating electrode* (RDE) discussed in Chapter 6 is an example of a device that uses forced convection to establish laminar flow across an electrode surface so that a well-defined steady-state diffusion layer is established. For the moment we will focus on diffusion as the main mechanism of mass transport. In terms of EIS, diffusion gives rise to characteristic *diffusion impedances* that are examined in more detail in the next section.

3.5 Diffusion: Fick's Laws

In electrochemistry, we encounter a new type of impedance element that is associated with the *diffusion of ions* to and from an electrode when an electrode reaction is taking place. For example, the oxidation of $Fe(CN)_6^{4-}$ ions to $Fe(CN)_6^{3-}$ ions involves the diffusion of the Fe(II) species to the electrode surface and the diffusion of the Fe(III) species back to the bulk of the solution. If the electron-transfer process is sufficiently fast, the overall rate of the process is determined by how fast the ions arrive at the electrode surface by diffusion. The reaction is said to be *diffusion-controlled*. This situation is discussed in detail in standard electrochemistry text books – see *Further Reading*. For the moment we will just state that changing the electrode potential perturbs the concentration gradients of redox species and so alters the diffusion-limited current. To obtain expressions for impedance elements associated with diffusion, we start with *Fick's laws of diffusion*, named after Adolf Fick, who started off as a physicist and mathematician but later switched to medicine. He was the first to measure cardiac output, and he applied his understanding of gas diffusion to good effect in physiology.

Fick's first law in one dimension looks very straightforward.

$$J(x) = -D\frac{dC(x)}{dx} \tag{3.7}$$

Here, $J(x)$ is the *flux* of material passing through a reference plane at the position x. The flux has units $mol\,cm^{-2}\,s^{-1}$. The $dC(x)/dx$ term is the *gradient of concentration* at x. Finally, D is the *diffusion coefficient* of the substance that is diffusing. Why the minus sign? At the molecular level, diffusion is a random process that results in the transport of material from regions of high concentration to regions of low concentration. Or to put it another way, substances diffuse down the concentration gradient. In electrochemistry, concentration gradients generally arise from the consumption or generation of materials in electrode reactions. Consumption of a reactant as the result of electron transfer at an electrode surface lowers its local concentration close to the surface. At the same time, generation of the product increases its concentration near the surface. This local perturbation of concentrations at $x = 0$ leads to the formation of concentration gradients that move reactants towards the electrode, and products, away. Reactant diffusion is towards lower values of x, and hence the flux is negative. Product diffusion takes place to higher values of x and so the flux is positive.

At this point, let us suppose that instead of applying a small sinusoidal voltage modulation we apply a *large amplitude potential step*. Furthermore, we will suppose that our solution contains only species R in addition to a large excess of supporting electrolyte. The potential is stepped from a potential where no oxidation of R occurs to a much more positive potential where R is very rapidly oxidized to O, almost completely depleting the concentration of R immediately adjacent to the electrode surface. If we take a 'snapshot' of the concentration gradients of R and O at some time after application of the potential step, they appear as illustrated in Figure 3.3. The important gradients as far as the current is concerned are those at the electrode surface ($x = 0$ in Figure 3.3). To convert from the flux of R at $x = 0$ to the current density, we need to know how much charge is passed when the reactant undergoes the electrode reaction. To convert from flux ($mol\,cm^{-2}\,s^{-1}$) to current density ($C\,s^{-1}\,cm^{-2} \equiv A\,cm^{-2}$), we multiply the flux by nF, where n is the number of electrons transferred in the elementary reaction, and F is the Faraday constant ($96{,}485\,C\,mol^{-1}$: $F = N_A q$ where N_A is the Avogadro constant, and q is the elementary charge).

For the anodic reaction $R = O + ne^-$, Fick's First Law then takes the form

$$j = nFD_R \left.\frac{dC_R(x)}{dx}\right|_{x=0} = -nFD_O \left.\frac{dC_O(x)}{dx}\right|_{x=0} \tag{3.8}$$

You will have noticed that the concentration gradient of R in Figure 3.3 decreases continually as we go further away from the electrode. This must

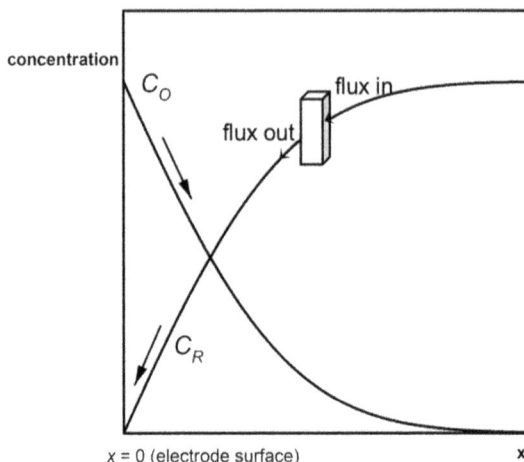

Figure 3.3. Typical concentration gradients when an electrode reaction is taking place, in this case, the oxidation of species R to species O takes place at potentials well positive of the standard reduction potential. The current density is determined by the concentration gradients at $x = 0$. An anodic current is taken by convention to be positive, and a cathodic current, negative. The decrease in the concentration gradient of R as we move away from the electrode means that the flux into the small reference volume shown is smaller than the flux out of it. It follows the concentration of R in the reference volume must fall with time. This is the basis for Fick's second law.

mean that the flux of R towards the surface also decreases with distance from the electrode surface, so that if we consider a small volume $A\,dx$ at any point x on the concentration curve for the species R diffusing towards the electrode, there is a lower flux into the volume at position $x + dx$ than out at x (see Figure 3.3). This must mean that the concentration in the reference volume falls with time. This applies at every point on a concentration curve for R so that the profiles are time dependent as shown in Figure 3.4. The opposite is true for the concentration profile of O, so the concentration of O at any point on the O curve will increase with time. If the small reference volume on the R concentration profile has an area $A = 1$ cm^2 and thickness dx, the change in concentration with time must be given by the number of moles entering per second minus the number of moles leaving per second divided by the volume of the thin box, which is $A\,dx = dx$. Using Fick's first law, we can write this for any point x on the concentration profiles as follows.

$$\frac{dC(x)}{dt} = \frac{-D\left.\frac{dc}{dx}\right|_{x+dx} - \left[-D\left.\frac{dc}{dx}\right|_{x}\right]}{dx} = \frac{D\left.\frac{dc}{dx}\right|_{x} - D\left.\frac{dc}{dx}\right|_{x+dx}}{dx} \qquad (3.9\text{a})$$

$$\frac{dC(x)}{dt} = D\frac{d^2 C(x)}{dx^2} \qquad (3.9\text{b})$$

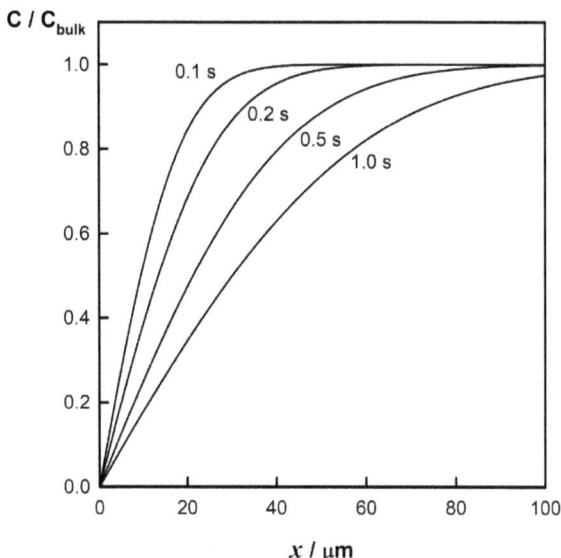

Figure 3.4. Evolution of concentration profiles with time shown obtained by solving Fick's second law using a diffusion coefficient $D = 10^{-5} \text{cm}^2\,\text{s}^{-1}$. $C_O^\infty = C_R^\infty$.

Fick's second law (equation (3.9b)) tells us that the gradient of concentration at any point x and time t depends on the second derivative of concentration with respect to distance x. We will not go into details about how we solve Fick's second law using Laplace transformation, but it is worth looking at the solution, which describes the way that the diffusion profiles of R and O change with time after application of a potential step (see Figure 3.4). The next theory note provides a brief explanation.

Theory Note 3.2. Fick's Second Law and Diffusion Profiles

When we solve Fick's second law with appropriate boundary conditions, we find that the concentration profiles of the reduced species as a function of time and distance are given by

$$C_R(x,t) = C_R^\infty \text{erf}\left[\frac{x}{2(D_R t)^{1/2}}\right] \tag{3.10}$$

where C_R^∞ is the bulk concentration of R and erf is the *Gaussian error function* (yes, our old friend Gauss again). The error function for a variable z is defined by

$$\text{erf}(z) = \frac{2}{\sqrt{\pi}} \int_0^z e^{-u^2} du \tag{3.11}$$

(Continued)

Theory Note 3.2. (*Continued*)

(Usually, u is replaced by t in this equation, but u is used here to avoid confusion with t for time. $\mathrm{erf}(z)$ is closely related to the familiar bell-shaped Gaussian probability curve.) Figure 3.5 illustrates the error function curve for negative and positive values of z.

We are interested in the slope of the profiles at $x = 0$, so noting from equation (3.11) that

$$\frac{d}{dz}[\mathrm{erf}(z)] = \frac{2}{\sqrt{\pi}}e^{-z^2} \tag{3.12}$$

where $z = x/[2(D_R t)^{1/2}]$, we find that the slope dC_R/dx at any point x and at time t is given by

$$\frac{dC_R(x,t)}{dx} = \frac{C_R^\infty}{\sqrt{\pi D_R t}}e^{-\frac{x^2}{4D_R t}} \tag{3.13}$$

We only require the slope at $x = 0$, so the exponential term becomes equal to 1 and we obtain

$$\frac{dC_R(x = 0,t)}{dx} = \frac{C_R^\infty}{\sqrt{\pi D_R t}} \tag{3.14}$$

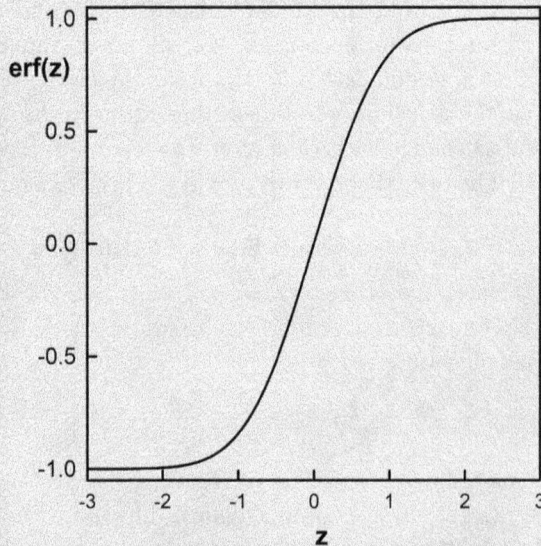

Figure 3.5 Plot of the Gaussian effort function $\mathrm{erf}(z)$ for positive and negative values of z.

Theory Note 3.2. (*Continued*)

Now we can calculate the current density using Fick's first law.

$$j(t) = nFD_R \left. \frac{dC_R(x,t)}{dx} \right|_{x=0} = \frac{nFD_R^{1/2}C_R^\infty}{\pi^{1/2}t^{1/2}} \qquad (3.15)$$

Equation (3.15) is known as the *Cottrell equation*, named after the American physical chemist and philanthropist Frederick Gardner Cottrell, who was a student of both van't Hoff and Ostwald. The equation predicts that the current response to a large amplitude response should decay with $1/\sqrt{t}$ as shown in Figure 3.6.

The dependence of the current on the inverse *square root* of time is mirrored in the inverse square root frequency dependence of the *Warburg diffusion impedance*, which we meet in the next section. In Chapter 6, we shall see that the limiting diffusion-controlled current at a rotating disc electrode depends on the square root of the rotation rate current.

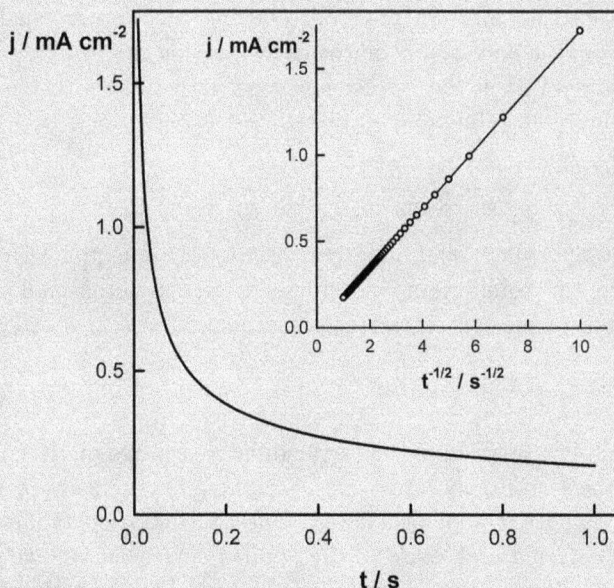

Figure 3.6 Large amplitude potential step experiment. Current density transient predicted by the Cottrell equation for $C_R = 10^{-6}$ mol cm^{-3} and $D_R = 10^{-5}$ cm^2 s^{-1}(1 mM). The inset shows a plot of current density vs. $1/\sqrt{t}$ that can be used to determine the diffusion coefficient.

Solutions of Fick's first and second laws are an essential part of the derivation of diffusion impedances for different circumstances. The two laws can be formulated in different co-ordinates depending on the geometry of the system. For example, we use spherical coordinates for hemispherical diffusion towards a very small electrode known as an ultramicroelectrode. We return to Fick's laws in Chapter 6 where we discuss experimental results obtained with a rotating disc electrode and with an ultramicroelectrode.

3.6 Diffusion Impedance: The 'Warburg'

So much for the large amplitude potential step. When we come to defining the diffusion impedance, we need to know how a *small amplitude* ac voltage perturbs the diffusion-controlled current. We will take the example of an electron transfer reaction that is sufficiently fast to ensure that (at all practical frequencies at least) the ratio of oxidized and reduced species at the surface is determined by the electrode potential E according to the Nernst equation – equation (3.2b). For simplicity, we will assume that we have equal bulk concentrations of O and R so that we have a well-established equilibrium potential obeying the Nernst equation. The dependence of the potential E on the concentration can be found by taking the derivative of E in the Nernst equation with respect to C_R and C_O, which, noting that $\frac{d}{dx}\ln(ax) = \frac{1}{x}$, gives

$$\frac{dE}{dC_O} = \frac{RT}{nFC_O} \qquad \frac{dE}{dC_R} = -\frac{RT}{nFC_R} \qquad (3.16)$$

It follows that a small sinusoidal perturbation of potential $\delta E e^{i\omega t}$ superimposed on the equilibrium potential will give rise to small sinusoidal perturbations of the concentrations of O and R at $x = 0$.

$$\frac{\delta C_O}{C_O} = \frac{nF}{RT}\delta E e^{i\omega t} \qquad \frac{\delta C_R}{C_R} = -\frac{nF}{RT}\delta E e^{i\omega t} \qquad (3.17)$$

When we apply a small amplitude ac modulation centred on the equilibrium potential, the surface concentrations C_O^0 and C_R^0 fluctuate above and below their bulk values by a small amount. The concentration profiles of both species vary during each cycle of the applied sinusoidal voltage as shown in Figure 3.7 for species O (the variation for R is essentially just the mirror image). As the figure shows, the perturbation of the concentration for this frequency (which is equivalent to 0.16 Hz) dies out over about 200 microns. This distance is equivalent to the *diffusion layer thickness*, which is inversely proportional to $\sqrt{\omega}$. The expression describing the variation of concentration with distance x and time t obtained by solution

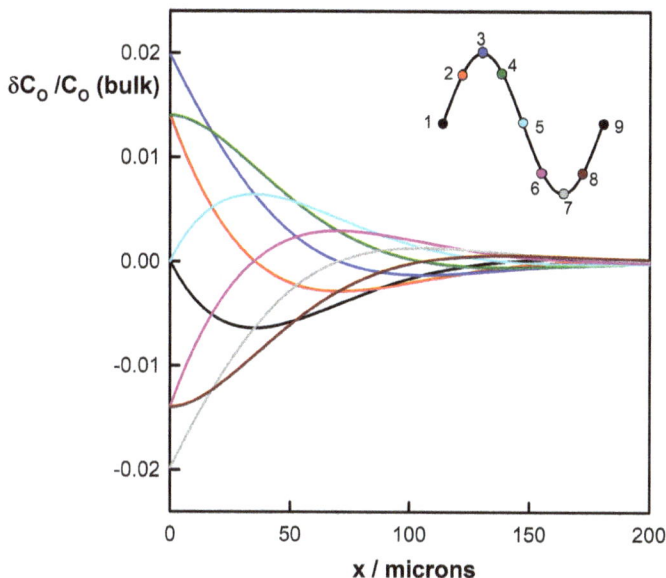

Figure 3.7. Normalized concentration profiles $\frac{\delta C_O}{C_O^\infty}$ of species O as a function of time caused by application of a 1 mV p-p sinusoidal potential modulation ($\omega = 1\mathrm{rad\,s}^{-1}$ (0.16 Hz), $C_O = C_R$, $D_O = D_R = 10^{-5}\mathrm{cm^2\,s}^{-1}$). δ_{ac}, the width of the diffusion layer here is around 200 microns. δ_{ac} depends on the $1/\sqrt{\omega}$. The different coloured concentration profiles correspond to he numbered points on the sinusoidal voltage modulation.

of Fick's second law for $C_O = C_R$ and $D_O = D_R = D$ is

$$\frac{\delta C_O(x,t)}{C_O^\infty} = \frac{nF\delta E}{2RT} \cdot e^{-\sqrt{\frac{\omega}{2D}}x} \cdot \sin\left(\omega t - -\sqrt{\frac{\omega}{2D}}x\right) \tag{3.18}$$

where δE is the peak-to-peak (p-p) amplitude of the modulation.

The small periodic voltage perturbation of the concentrations at $x = 0$ and the associated current response described by Fick's laws define a *diffusion impedance*. If diffusion takes place from a large volume of solution to a flat electrode (*semi-infinite linear diffusion*), the diffusion impedance is called the *Warburg impedance*. The Warburg impedance is interesting because it has *equal real and imaginary components at all frequencies*. As if that was not interesting enough, it also *depends on the inverse square root of frequency*.

$$Z_W = \sigma\omega^{-1/2} - j\sigma\omega^{-1/2} \tag{3.19}$$

σ is a constant that depends on the concentrations (C_O, C_R) and diffusion coefficients (D_O, D_R) of the oxidized and reduced species. Z_W is clearly

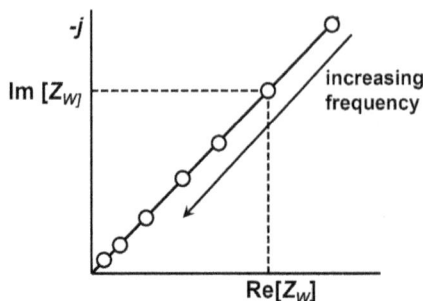

Figure 3.8. The Warburg impedance for semi-infinite linear diffusion. The constant phase shift is $-45°$ $(-\pi/4)$.

a new kind of *constant phase shift element* that obeys the relationship $|Z_W| \propto f^{-1/2}$ and which has a *constant phase shift of* $-45°(-\pi/4)$. The Nyquist plot of Z_W must therefore have the shape shown in Figure 3.8.

The constant σ in equation (3.19) is given by

$$\sigma = \frac{RT}{\sqrt{2}An^2F^2}\left(\frac{1}{D_O^{1/2}C_O} + \frac{1}{D_R^{1/2}C_R}\right) \tag{3.20}$$

To give an idea of the value of σ, we can put in values for the $Fe(CN)_6^{4-}/Fe(CN)_6^{3-}$ reaction. Area $A = 1$ cm^2; number of electrons transferred $(n) = 1$; R, T and F – standard values; concentrations C_O and C_R, both 10^{-5} mol cm^{-3} (note that the concentration units mol cm^{-3} are consistent with those of the diffusion coefficient: cm^2 s^{-1}); diffusion coefficients: $D_O \approx D_R = 10^{-5}$ cm^2 s^{-1}. This gives $\sigma = 11.9$. What about the units of σ? They must be $\Omega\,\text{rad}^{1/2}\,\text{s}^{-1/2}$ to satisfy equation (3.19).

How can we simulate a Warburg impedance in ZView®? Unfortunately, the two types of Warburg impedances that we can find in the menu for the Equivalent Circuit tool are not suitable (more of these Warburg elements later). Instead, we use the *constant phase shift element* (CPE) listed as CPE#1 in the menu, as shown in Figure 3.9.

But what exactly is this constant phase shift element? We should check the help menu in ZView® to find out before we start using it. Figure 3.10 shows the information that the ZView® Help file provides.

Does this look suitable to model Z_W? T is a constant, i is our j and w is our ω. ^P means raised to the power of P, so if we let $P = 0.5$, then the CPE#1 is as follows:

$$Z = \frac{1}{T(j\omega)^{\frac{1}{2}}} \tag{3.21}$$

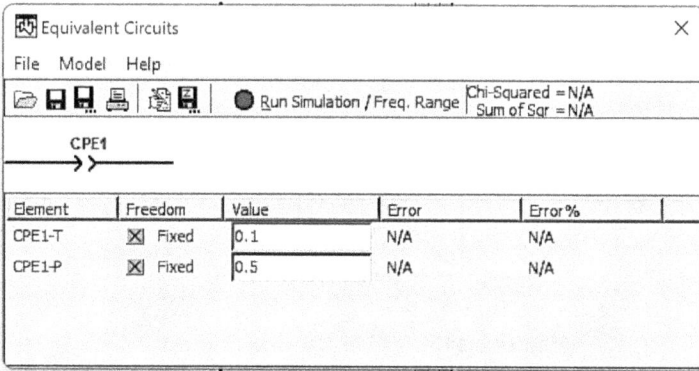

Figure 3.9. The constant phase shift element CPE#1 in ZView®.

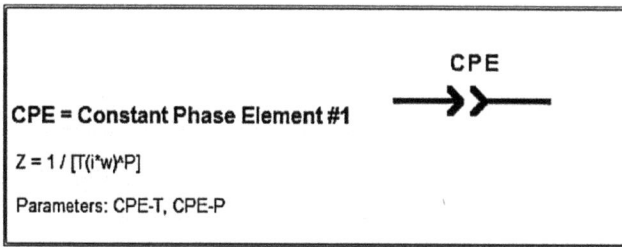

Figure 3.10. ZView® information about the constant phase shift element (CPE).

This doesn't look at all like equation (3.19), which was $Z_W = \sigma\omega^{1/2} - j\sigma\omega^{-1/2}$.

The $\omega^{-1/2}$ bit looks OK, but the expression is also asking us to take *the square root of j*. How can we do this? Time for another Theory Note.

Theory Note 3.3. Taking the Square Root of j

Time to appeal to Herr Euler again for inspiration. Remember his equation?

$$e^{i\varphi} = \cos\varphi + j\sin\varphi \qquad (3.22)$$

What happens if the angle $\varphi = \pi/2$ (i.e., 90°)?

$$e^{j\frac{\pi}{2}} = \cos\frac{\pi}{2} + j\sin\frac{\pi}{2} = 0 + j(1) = j \qquad (3.23)$$

(Continued)

Theory Note 3.3. (*Continued*)

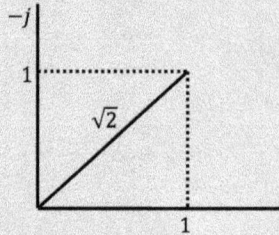

Figure 3.11. Finding the square root of j.

Neat! Now we know $j = e^{j\frac{\pi}{2}}$, we just need to take the square root of $e^{j\frac{\pi}{2}}$. This means we raise $e^{j\frac{\pi}{2}}$ to the power $1/2$. The answer is $e^{j\frac{\pi}{4}}$ because $(e^x)^a = e^{ax}$.

Now we use Euler again to convert $e^{j\frac{\pi}{4}}$ back to a complex number. We obtain

$$\sqrt{j} = e^{j\frac{\pi}{4}} = \cos\frac{\pi}{4} + j\sin\frac{\pi}{4} = \frac{1}{\sqrt{2}} + j\frac{1}{\sqrt{2}} \tag{3.24}$$

The triangle in Figure 3.11 shows where the $1/\sqrt{2}$ terms in equation (3.24) come from. Both the sine and cosine of $\pi/4(45°)$ are equal to $1/\sqrt{2}$.

What was the point of all that? We just found that *the square root of j has equal real and imaginary components and a phase angle of $45°(\pi/4)$* – just what we need for the Warburg diffusion impedance. We can now finish the job (using the complex conjugate again).

$$Z = \frac{1}{T(j\omega)^{\frac{1}{2}}} = \frac{\omega^{-\frac{1}{2}}}{T\sqrt{j}} = \frac{\omega^{-\frac{1}{2}}}{T\left(\frac{1}{\sqrt{2}} + j\frac{1}{\sqrt{2}}\right)} = \frac{\left(\frac{1}{\sqrt{2}} - j\frac{1}{\sqrt{2}}\right)}{T\left(\frac{1}{2} + \frac{1}{2}\right)}\omega^{-\frac{1}{2}}$$

$$Z = \frac{1}{T\sqrt{2}}\omega^{-1/2} - j\frac{1}{T\sqrt{2}}\omega^{-1/2} \tag{3.25}$$

We just showed that σ in the equation for the Warburg impedance is the same as $\frac{1}{T\sqrt{2}}$, i.e., for the simulation of the Warburg impedance in ZView®, we just need to use $P = 0.5$ and $T = \frac{1}{\sigma\sqrt{2}}$.

So, let us run the ZView® simulation with $P = 0.5$, and since $\sigma = \frac{1}{T\sqrt{2}}$, we choose a value of $T = 0.074$, which corresponds to $\sigma = 11.9$. Figure 3.12 shows the simulation result.

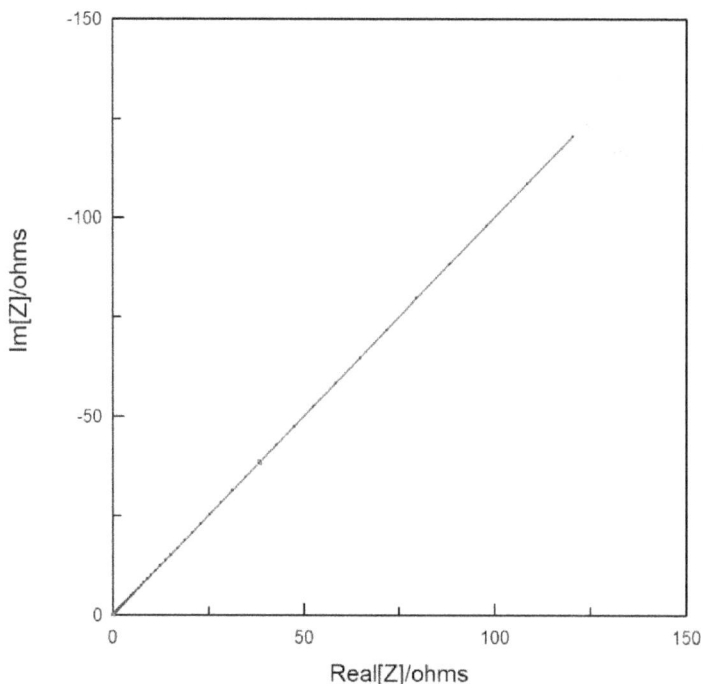

Figure 3.12. The Warburg impedance calculated for $T = 0.074$, $P = 0.5$. These values correspond to the Warburg impedance with $\sigma = 11.9$ (calculated for 10 mM $\mathrm{Fe(CN)_6^{4-}}/\mathrm{Fe(CN)_6^{3-}}$).

The tell-tale signs of the Warburg impedance are the $-45°$ line in the Nyquist plot, the slope of -0.5 (corresponding to an $\omega^{-1/2}$ frequency dependence) in the Bode $\log_{10}|Z|$, $\log_{10} f$ plot and the constant angle of $-45°$ in the phase plot.

3.7 Putting It All Together: The Randles Equivalent Circuit

Now we are ready to describe an electrode process that is controlled by diffusion and by the kinetics of electron transfer. The equivalent circuit first proposed as long ago as 1947 by the brilliant Birmingham-based English electrochemist *John Randles* (as a PhD student, I vividly remember sitting next to John Randles as he evaluated an integral on a small scrap of paper before making an incisive comment during a lecture by the eminent US electrochemist Ernest Yeager at Southampton University). The Randles circuit contains the three circuit elements that we have looked at so far in this chapter: the double-layer capacitance, the Faradaic resistance and the semi-infinite Warburg impedance. It also includes the series resistance of the electrolyte (Figure 3.13).

Now let us use the same values of C_O, C_R (10^{-5} mol cm^{-3}) and D (10^{-5} cm^2 s^{-1}) for the $Fe(CN)_6^{4-}/Fe(CN)_6^{3-}$ reaction. In addition, we will specify $\alpha = 0.5$ and $k^\circ = 10^{-2}$ cm s^{-1}, giving an exchange current density j_0 of around 10^{-2} A cm^{-2} and a corresponding Faradaic resistance

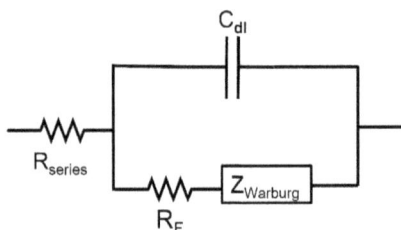

Figure 3.13.   The Randles equivalent circuit for an electrode reaction.

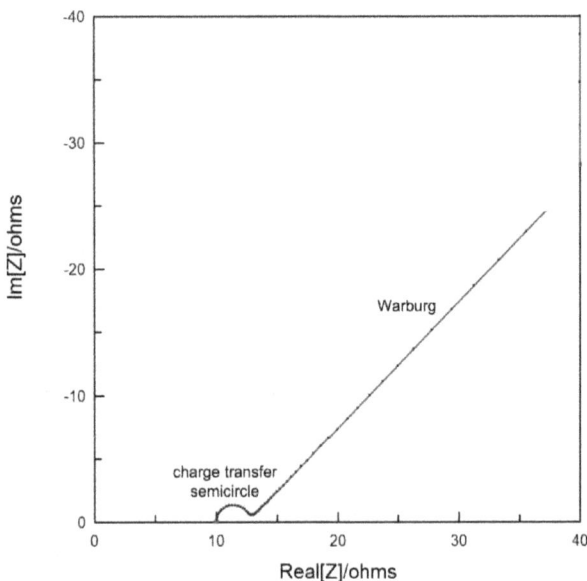

Figure 3.14.   Nyquist and Bode impedance plots for the Randles equivalent circuit. $R_{series} = 10\,\Omega$, $R_F = 2.6\,\Omega$, $C_{dl} = 30\,\mu F$. Warburg simulated by a CPE#1 with $T = 0.074$, $P = 0.5$.

$R_F = 2.6\,\Omega\,\text{cm}^2$. Running ZView® with the Randles circuit and replacing the Warburg impedance by a CPE#1 as before gives the result shown in Figure 3.14 for an electrode area of 1 cm².

This result illustrates the power of EIS to deconvolute the parameters for a quite complex process involving double-layer charging, diffusion and charge transfer.

3.8 More ZView® Circuit Elements: The Finite Length Warburg (Short Circuit Terminus)

If you use ZView®, you will be amazed at the number of circuit elements that are available. Here, we will look at another one to illustrate the need to understand exactly what we are dealing with when we choose one of these elements for fitting our experimental data. The element we will look at next is *Finite Length Warburg*. Figure 3.15 shows the information about it in the ZView® help file.

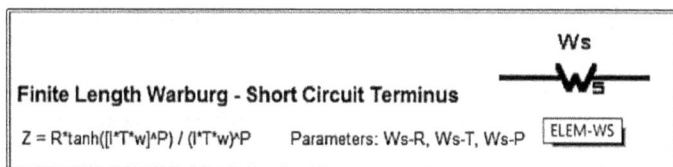

Ws

Finite Length Warburg - Short Circuit Terminus

$Z = R^*\tanh([I^*T^*w]^\wedge P) / (I^*T^*w)^\wedge P$ Parameters: Ws-R, Ws-T, Ws-P   ELEM-WS

Figure 3.15. ZView® Help File on the Finite Length Warburg.

It looks complicated, but the first question to ask is what physical process does this element represent? The answer to this question is suggested by the name of the element. In the previous section, we discussed the Warburg impedance for semi-infinite linear diffusion. In the case of the finite Warburg, we can guess that diffusion is taking place within a confined space. Let us look at two examples of this. As an example, we will use the reaction

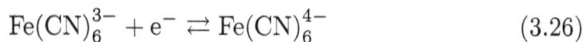

$$\text{Fe(CN)}_6^{3-} + e^- \rightleftarrows \text{Fe(CN)}_6^{4-} \tag{3.26}$$

For simplicity, we will assume that the concentrations of the oxidized and reduced ions in the solution are the same. We met this redox reaction already when discussing the semi-infinite Warburg element. In that case, the reaction is initially at equilibrium, and then the system is perturbed by a small ac signal that is centred on the equilibrium potential. In this case, no dc current flows. However, we might be interested in studying the reaction at a potential that is either positive or negative of the equilibrium potential. We need to have a *steady-state* dc current before we can perturb the system with our small ac voltage. This is a problem in an unstirred solution because the diffusion-limited current is time dependent as we showed earlier, so we do not have a steady state. To establish *steady-state mass transfer conditions*, we need to take a different approach. One way of doing this is to use a *rotating disc electrode*. This is a rotating cylinder with a circular disc electrode in its end face as shown in Figure 3.16. The rotating electrode acts as a pump, pulling liquid up towards the electrode and spinning it out tangentially. This sets up a reproducible hydrodynamic condition that defines an unstirred boundary layer of thickness δ close to the electrode surface. Beyond this layer, the solution is stirred so that the concentrations of oxidized and reduced redox species are constant. Thanks to the distinguished physical chemist and Soviet dissident Benjamin Levich, we know that the thickness, δ, of the boundary layer is inversely proportional to the square root of the rotation rate. Levich solved many hydrodynamic problems, including the rotating disc electrode, and his book *Physiochemical Hydrodynamics* is the most important text on this topic. A prominent *refusenik*, Levich had to wait 6 years before being allowed to leave the Soviet Union in 1978 to work in the USA and Israel.

Figure 3.16 illustrates the situation where the concentrations of oxidized (O) and reduced (R) species are the same in the bulk of the solution, and

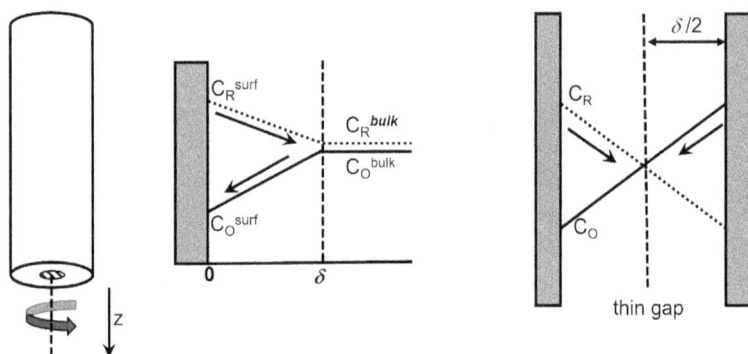

Figure 3.16. The rotating disc electrode (left) sets up a well-defined boundary layer in which diffusion of species to and from the electrode takes place. As the potential is perturbed on either side of the equilibrium potential, the gradients for O and R swap over, and ions diffuse in the opposite direction (the gradients have been greatly exaggerated here so they are visible). The thin layer cell (right) is similar, except that the gradients cross in the centre of the gap, so that δ in the expressions for the finite Warburg impedance should be replaced by $\delta/2$.

the electrode potential is held slightly negative of the equilibrium potential so that net reduction of $Fe(CN)_6^{3-}$ to $Fe(CN)_6^{4-}$ takes place, and a cathodic current flows. The concentration profiles in the boundary layer are linear, so there is a steady-state dc current that depends inversely on δ, and hence on the square root of the rotation rate (the profiles have been greatly exaggerated in the figure for clarity). Now if we change the potential to slightly positive of the equilibrium potential, the situation will reverse, and net oxidation of $Fe(CN)_6^{4-}$ to $Fe(CN)_6^{3-}$ will take place with an anodic current flowing. The concentration profiles for O and R shown in the figure will then swap places, and the ions will diffuse in the opposite direction. In both cases, the ratio of the concentrations of the oxidized and reduced species *at the surface of the electrode* is determined by the Nernst equation, equation (3.2b), if the reaction is diffusion controlled. At the same time, the stirring action of the electrode fixes the concentrations of O and R at their bulk values at $z = \delta$ (the coordinate z is used here since x, y are the coordinates in the plane of the electrode and the z coordinate is perpendicular to the electrode surface). Diffusion is therefore restricted to the region between $z = 0$ and $z = \delta$, hence *the finite length Warburg.*

Another example of the finite Warburg impedance is a thin-layer cell, where two flat electrodes are separated by a small electrolyte-filled gap. When a redox species is oxidized at one electrode, it diffuses across the gap to the other electrode where it is reduced (i.e., recycled). Under steady-state conditions, the concentration gradients are linear as shown in Figure 3.16.

In the impedance measurements on these systems, we can perturb the electrode by a small *ac* signal centred on *any* potential and measure the resulting *ac* current, which is centred about the steady-state (dc) current. It is no longer necessary to restrict measurements to *ac* perturbation centred on the equilibrium potential. A nice application of the thin-layer cell in a practical device is the dye-sensitized solar cell discussed in Chapter 7.

Now we know what physical situation is described by the finite Warburg, what about the impedance? Solution of the diffusion problem for the rotating disc case gives the following equation for $Z_{W,s}$ (for the thin-layer cell, δ is replaced by $\delta/2$).

$$Z_{W,s} = \frac{Z_{W,s}(0)\tanh(j\Lambda)^{0.5}}{(j\Lambda)^{0.5}}$$

where

$$Z_{W,s}(0) = \frac{2RT\delta}{An^2F^2CD} \quad \text{and} \quad \Lambda = \frac{\omega\delta^2}{D} \tag{3.27}$$

Here, D is the diffusion coefficient of the redox species (assumed to be the same for both oxidized and reduced species), and δ is the width of the diffusion layer shown in Figure 3.16. Since the units of the diffusion coefficient, D, are $cm^2\ s^{-1}$, you can see that Λ *is a dimensionless parameter.*

Looking back at the equation in the ZView® help file (Figure 3.15), we see a tanh term. tanh is a *hyperbolic function* (hence the h at the end). Since panic may be setting in at this point, it is time to take a deep breath for a Theory Note.

Theory Note 3.4. Hyperbolic Functions

In the solution of the mathematical problems (*differential equations*) associated with physical processes, we often come across *exponential functions.* We shall not go into details of the origin of the hyperbolic functions here. All you need to know is that they are a form of shorthand for particular exponential functions. Here are the definitions of $\sinh x$, $\cosh x$ and $\tanh x$.

Hyperbolic sine: sinh.

$$\sinh x = \frac{e^x - e^{-x}}{2} \tag{3.28}$$

Hyperbolic cosine: cosh.

$$\cosh x = \frac{e^x + e^{-x}}{2} \tag{3.29}$$

Hyperbolic tangent: tanh.

$$\tanh x = \frac{\sinh x}{\cosh x} = \frac{e^x - e^{-x}}{e^x + e^{-x}} \tag{3.30}$$

It is helpful to look at what happens to $\tanh x$ when x gets small, since as we shall see in what follows, this is important for the low-frequency limit of the impedance. Remember that e^x can be replaced by $1+x$ when x is very small. Have a look back at Chapter 1, Theory Note 1. So, for $x \ll 1$.

$$\tanh x = \frac{e^x - e^{-x}}{e^x + e^{-x}} \rightarrow \frac{(1+x)-(1-x)}{(1+x)+(1-x)} = \frac{2x}{2} = x \tag{3.31}$$

What if x is large? Then e^x is much larger than e^{-x}, and so $\tanh x$ tends to 1.

Now it is time to compare equation (3.27) and the expression in the ZView® help file (Figure 3.15) to see if we can work out what the symbols mean. The ZView® help file tells us that the T in the expression is equal to L^2/D, i.e., $T = L^2/D$, where L is the diffusion distance and D is the diffusion coefficient. We can conclude that L in the formula is the same as δ in equation (3.26). P is obviously 0.5 and w is ω. Looking at the units, we see that T has units of time, so T multiplied by ω is dimensionless – this is the Λ in equation (3.15). This just leaves the mysterious 'l' in the ZView® equation. This should actually be j! So if we replace the 'l' by j, the two expressions are identical and obviously $R = Z_{W,s}(0)$.

The parameter T in the ZView® program corresponds to the time that ions take to diffuse across the diffusion layer from $x = \delta$ to $x = 0$ or vice versa. In our terminology, this *diffusion time constant* is given by

$$\tau_{\text{diff}} = \frac{\delta^2}{D} \tag{3.32}$$

Now, we can use our newly acquired knowledge about the tanh function to ask what happens to $Z_{W,s}$ as the frequency ω tends towards zero. Here is the expression for $Z_{W,s}$ again.

$$Z_{W,s} = \frac{Z_{W,s}(0) \tanh(j\Lambda)^{0.5}}{(j\Lambda)^{0.5}} \tag{3.33}$$

In ZView®

$$Z = R^* \tanh([l^* T^* w]^\wedge P)/(l^* T^* w)^\wedge P \tag{3.34}$$

Comparing the two expressions, you can see that the x in $\tanh(x)$ in our case is equal to $(j\Lambda)^{0.5}$. So the tanh term becomes $(j\Lambda)^{0.5}$, which then cancels out with the $(j\Lambda)^{0.5}$ in the denominator. This leaves the low-frequency limit equal to $Z_{W,s}$, which of course makes sense. Note that since the j terms have cancelled out, $Z_{W,s}$ *must be a real quantity*.

So let us put in some values to obtain $Z_{W,s}(0)$. Using $\delta = 10^{-2}$ cm, $A = 1$ cm^2, $n = 1$, $C_O = C_R = C = 10^{-5}$ mol cm^{-3}, $D = 10^{-5}$ cm^2 s^{-1} gives $Z_{W,s}(0) = 53.2\,\Omega$. We can now use this value in the ZView® program. We also need T. It is equal to $\delta^2/D = 10^{-4}$ cm$^2/10^{-5}$ cm^2 s$^{-1} = 10$ s. The result of the impedance calculation is shown in Figure 3.17.

The impedance response at high frequencies is a straight line on the Nyquist plot with an angle of $-45°$, and this corresponds to the semi-infinite Warburg impedance. As the frequency decreases, the plot deviates from the line and loops round into a semicircle that intercepts the real axis at the zero-frequency intercept $Z_{W,s}(0)$.

This example illustrates the importance of understanding what the parameters mean when using ZView® to model impedance responses or, more importantly, to fit experimental data. It is tempting just to choose some values and then to run the fit without appreciating what the

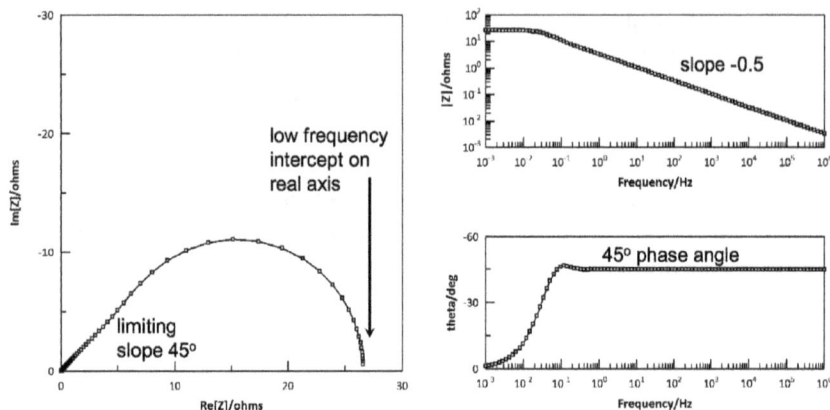

Figure 3.17. Finite Warburg impedance calculated for a rotating disc electrode with a boundary layer thickness of $\delta = 10^{-2}$ cm, $C_O = C_R = 10^{-5}$ mol cm^{-3}, $D_O = D_R = 10^{-5}$ cm^2 s^{-1}. The corresponding values used in the ZView® modelling are $R = 26.6$, $T = 10$, $P = 0.5$.

parameters T and R mean. Even more dangerous is the temptation to run a fit with P free to take any value. Only the value $P = 0.5$ has any meaning in the context of the finite Warburg impedance, and *it should therefore be fixed during the fitting process.*

3.9 Another Warburg: The Finite Length Warburg Open Circuit Terminus

You will also find this element in the ZView® menu. What does it describe? In the previous cases of the rotating disc electrode and thin-layer cell, we saw that we could get a steady-state dc current. In some cases, we can have a situation where a current passes for some time and then stops. An example is charging the lithium-ion battery in your mobile phone. Another is the oxidation of a layer of a conducting polymer-like polyaniline or polypyrrole. In both cases, electron transfer is accompanied by ion transfer.

In the lithium-ion battery (Figure 3.18), we are dealing with *lithium intercalation* into the cathode material, which involves movement of lithium ions from the electrolyte into the cathode accompanied by transfer of electrons from the contact to the lithium ions. The process stops when intercalation is complete. Discharge of the battery involves the reverse process, with Li ions leaving the layer and electrons passing to the contact.

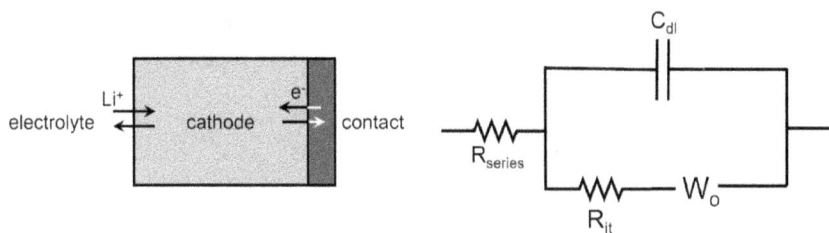

Figure 3.18. Intercalation cathode in a lithium-ion battery. During charging, Li^+ ions are transferred from the solution into the cathode material, and electrons are supplied from the back contact. The process is reversed on discharge. The equivalent circuit on the right contains an ion transfer resistance R_{it} and a double-layer capacitance associated with the electrolyte/cathode interface. Diffusion within the cathode layer is modelled by the finite Warburg element (open terminus).

The intercalation process is described by diffusion in a thin layer of cathode material. The cathode back contact forms a blocking contact for ions, hence *open circuit terminus*. In the case of conducting polymers, electrons move out of the material as it is oxidized, and negatively charged anions move in to preserve electroneutrality, but cannot enter the contact.

The finite Warburg impedance (open) shown in equation (3.35) has the form very similar to the one for the finite Warburg impedance (short). You can compare it with equation (3.27).

$$Z_{W,o} = R_W \frac{\coth(j\Lambda)^{0.5}}{(j\Lambda)^{0.5}} \quad \text{with} \quad \Lambda = \frac{\omega\delta^2}{D} \quad R_W = \frac{V_m \left(\frac{dE}{dy}\right)\delta}{AFD} \quad (3.35)$$

In this case, δ is the thickness of the cathode layer, and D is the diffusion coefficient of the Li^+ ions. R_W depends on the way the voltage changes with the mole fraction y of Li in the cathode, dE/dy, as well as on V_m, the molar volume of the cathode material and D, the diffusion coefficient of ions in the cathode. The only major difference compared with the expression for the finite Warburg (short) is that instead of a tanh term we now have a coth term.

You probably know that $\cotan(x) = 1/\tan(x)$. Similarly, $\coth(x) = 1/\tanh x$. Looking back at Theory Note 3.4, equation (3.30), we see that

$$\coth(x) = \frac{e^x + e^{-x}}{e^x - e^{-x}} \quad (3.36)$$

Using the approach illustrated in Theory Note 3.4, you can easily verify that for small x, $\coth(x)$ approaches $1/x$, whereas for large x, $\coth(x)$ approaches 1.

The *low-frequency limit* of the finite Warburg impedance in equation (3.35) is obtained by noting that $\coth(j\Lambda^{0.5})$ can be replaced by $1/(j\Lambda^{0.5})$.

$$Z_{W,o}|_{\omega\to 0} = R_W \frac{\frac{1}{(j\Lambda)^{0.5}}}{(j\Lambda)^{0.5}} = R_W\frac{1}{j\Lambda} = R_W\frac{D}{j\omega\delta^2} = -jR_W\frac{D}{\omega\delta^2} \quad (3.37)$$

This result shows that as the frequency approaches zero, the diffusion impedance becomes infinite and imaginary. In other words, it behaves just like a capacitance.

So, what about the *high-frequency limit*? Since $\coth(x)$ can be replaced by 1 for large x, we are just left with

$$Z_{W,o}|_{\omega\to\infty} = R_W\frac{1}{(j\Lambda)^{0.5}} = \frac{R_W}{\sqrt{\Lambda}}\frac{1}{\sqrt{j}} \quad (3.38)$$

There is the *square root of j* again – do you remember what it is? Look back at Theory Note 3.3 where we showed that $\sqrt{j} = \frac{1}{\sqrt{2}} + j\frac{1}{\sqrt{2}}$. Using this information, our impedance becomes

$$Z_{W,o}|_{\omega\to\infty} = \frac{R_W}{\sqrt{\Lambda}}\frac{1}{\left(\frac{1}{\sqrt{2}} + j\frac{1}{\sqrt{2}}\right)} = \frac{R_W}{\sqrt{\Lambda}}\frac{\sqrt{2}}{(1+j)} = \frac{R_W}{\sqrt{\Lambda}}\frac{\sqrt{2}(1-j)}{1+1}$$

$$= \frac{\sqrt{2}R_W}{2\sqrt{\Lambda}}(1-j) \quad (3.39)$$

Now we can replace Λ by $\Lambda = \frac{\omega\delta^2}{D}$ to obtain our final expression

$$Z_{W,o}|_{\omega\to\infty} = \frac{R_W\sqrt{2D}}{2\delta}\omega^{-0.5} - j\frac{R_W\sqrt{2D}}{2\delta}\omega^{-0.5} \quad (3.40)$$

So, we have *equal real and imaginary components* just like the semi-infinite Warburg. We therefore expect the high-frequency limit to be a straight line with a slope of $-45°$ in the Nyquist plot.

To summarize, we expect the high-frequency impedance to look like a normal Warburg impedance, but with decreasing frequencies, the Nyquist plot should deviate upwards to form a vertical line looking like the response of a capacitor.

Let us see what ZView® has to offer. $Z_{W,o}$ is defined in the help file as

$$Z = R^*\text{ctnh}([I^*T^*w]\hat{}\,P)/(I^*T^*w)\hat{}\,P \quad (3.41)$$

We recognize ctnh as coth. P must obviously be 0.5. I^*T^*w is $jT\omega$ with $T = \delta^2/D$. You should be able to see that the ZView® expression is the same as equation (3.35).

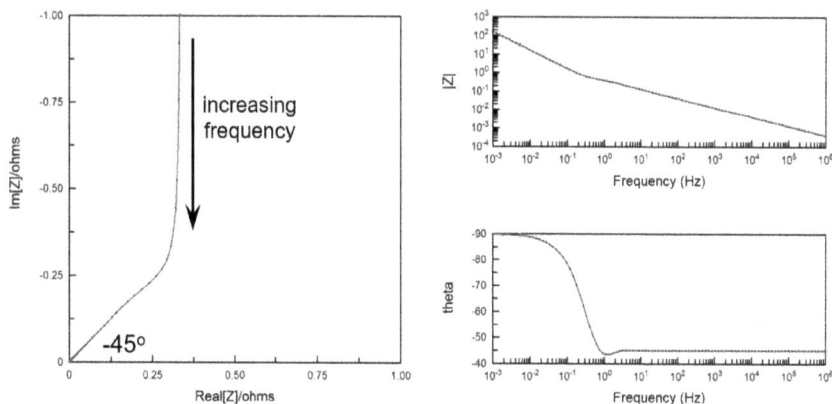

Figure 3.19. Nyquist and Bode plots for the finite Warburg impedance (open circuit termination). The impedances have been normalized to R_W, see equation (3.35), and the frequencies have been normalized to the frequency: $1/T = \delta^2/D$. Note the change in slope of the Bode plot from -1 at low frequencies (capacitor behaviour) to -0.5 at higher frequencies (Warburg behaviour).

To illustrate the finite Warburg (open), we can normalize the impedance to R_W and the frequency to the inverse of the diffusion time constant T (i.e., we multiply the angular frequencies by T). To do this in ZView®, we enter 1 for both R and T and set $P = 0.5$. The result, which is shown in Figure 3.19, confirms the limiting low- and high-frequency behaviours.

3.10 De Levie Finite Pore Impedance

So far, we have not paid much attention to the geometrical nature of the electrode in the electrochemical systems being studied by EIS. Since many practical applications of electrochemistry (batteries, supercapacitors) involve *porous electrodes*, we will now look at the *cylindrical pore model* associated with the distinguished Dutch/American electrochemist Robert de Levie, who first considered porous electrodes in his 1963 thesis *On porous electrodes* (University of Amsterdam). The idea is that an electrode contains a large number of pores, increasing its surface area while at the same time introducing a series resistance associated with the electrolyte filling the pores. The de Levie model considers a *cylindrical pore of finite length and fixed radius*. The series resistance is distributed from the top to the bottom of the pore and the impedance associated with the double layer and Faradaic reactions is likewise distributed down the internal area of the pore as illustrated in Figure 3.20. *Note that the resistances appear in series and the interfacial impedances appear in parallel.* The model makes the simplifying assumption that the voltage drop down the pore is sufficiently small that the Faradaic impedance is independent of position.

The resistance dR_{el} of the small volume element of length dx is equal to $\rho dx/\pi r^2$, where *ρ is the resistivity of the electrolyte* (units Ω cm) and r is the pore radius. The interfacial impedance dZ_{int} is equal to $Z_{int}/2\pi r dx$, where Z_{int} is the *interfacial impedance* in units Ω cm^2 (Z_{int} is assumed to be independent of position in the pore). The model is illustrated in Figure 3.20.

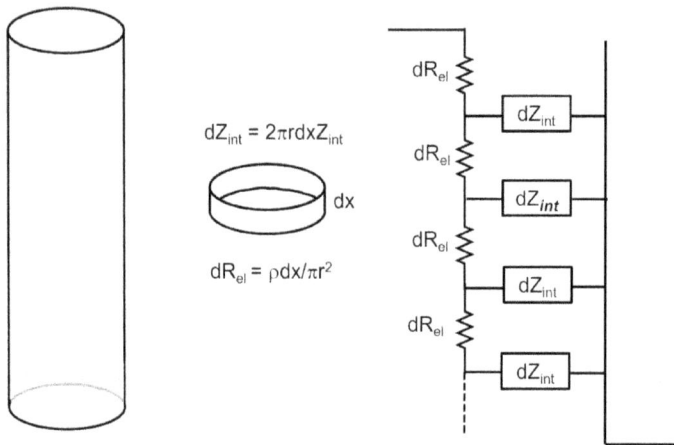

Figure 3.20.   de Levie cylindrical pore model. The electrolyte resistance is distributed down the pore to the bottom, and the interfacial impedance is distributed on the walls of the pore, giving the equivalent circuit shown on the right.

The impedance of a single pore of length l is given by

$$Z_{pore} = \frac{R_\Omega}{\sqrt{\Lambda}}\coth\left(\Lambda^{\frac{1}{2}}\right) \quad \text{where} \quad R_\Omega = \frac{\rho l}{\pi r^2} \quad \text{and} \quad \Lambda = \frac{\frac{2\rho l}{r}^2}{Z_{int}} \quad (3.42)$$

You can check that Λ is dimensionless (remember the units of Z_{int} are Ω cm^2).

We now have a closer look at a special case. If the interfacial impedance elements in Figure 3.20 are just due to the double-layer capacitance because there is no electrochemical reaction taking place, then we have an *ideally polarizable* porous electrode. In this case, the interfacial impedance is just $Z_{int} = -j/\omega C_{dl}$. The dimensionless parameter Λ is therefore

$$\Lambda = \frac{2\rho l^2}{r Z_{int}} = j\frac{2\rho l^2 \omega C_{dl}}{r} \quad (3.43)$$

Does this equation ring any bells? Look back at equation (3.35) for the finite Warburg (open circuit terminus). Here are the two equations again.

$$Z_{W,o} = R_W \frac{\coth(j\Lambda)^{0.5}}{(j\Lambda)^{0.5}} \quad \text{with} \quad \Lambda = \frac{\omega\delta^2}{D} \tag{3.44}$$

$$Z_{\text{pore}} = \frac{R_\Omega}{\Lambda^{\frac{1}{2}}}\coth\left(\Lambda^{\frac{1}{2}}\right) \quad \text{with} \quad \Lambda = \frac{\frac{2\rho l}{r}^2}{Z_{\text{int}}} \tag{3.45}$$

You can see that the finite Warburg (open terminus) had $j\Lambda$ terms rather than the Λ terms in the finite pore model, but the form is the same. You can also see that the special case of the ideally polarizable pore (just the double-layer capacitance) introduces the j that makes the equations identical. Only the meaning of the constants in the two cases are quite different. This means that *we can represent finite diffusion by an equivalent circuit like the one for the cylindrical pore.* As you might have guessed, semi-infinite diffusion can be represented by a semi-infinite pore ($l \to \infty$). This highlights the important conclusion that *quite different physical systems may give the same kind of impedance response.*

Now we can take a closer look at the *low-frequency limit* of the pore impedance using our 'common-sense' approach. The current has a choice at each 'T-junction' (or *node*) in the network shown in Figure 3.20. It can either carry on down the pore through the resistance or it can flow to the wall of the pore through the interfacial impedance. Each node therefore acts as a *current divider*. This means that as we go further and further down the pore, the current gets smaller and smaller. Eventually, the current will just fizzle out unless the bottom of the pore is reached before that happens. If we are perturbing the system, the same picture is valid. The effects of the perturbation will get smaller and smaller as we go down the pore. We can think of our perturbation penetrating some distance down the pore. This distance – the *penetration depth* – may be smaller than the pore length, so we don't get any information about the bottom of the pore. We can put this on a more exact basis in another of our *Theory Notes.*

Theory Note 3.5. Defining the Penetration Depth for a Porous Electrode

We are considering the case where the interfacial impedance is a parallel combination of the Faradaic resistance R_F and the double-layer capacitance C_{dl}. If we go to very low frequencies, then we can forget about the double-layer capacitance since its impedance is much larger

(*Continued*)

Theory Note 3.5. (*Continued*)

than R_F. The impedance of the pore must therefore become real under these conditions, and $Z_{\text{int}} = R_F$. Looking back at the definition of Λ, we can now obtain our low-frequency limit of Λ.

$$\Lambda\omega \to 0 = \frac{\frac{2\rho l^2}{r}}{Z_{\text{int}}} = \frac{2\rho l^2}{rR_F} \tag{3.46}$$

The *low-frequency limit of the pore impedance* is therefore

$$Z_{\text{pore}}|_{\omega\to 0} = \sqrt{\frac{\rho R_F}{2\pi^2 r^3}}\coth\left(l\sqrt{\frac{2\rho}{rR_F}}\right) \tag{3.47}$$

It is time to use what we know about how $\coth(x)$ behaves for small and very large values of x. This will depend on how large the square root part of the coth term is compared with l, the length of the pore. Since l is a distance, the *inverse of the square root term* must also be a distance, since Λ is dimensionless. Let us call this distance λ. Here it is.

$$\lambda = \sqrt{\frac{rR_F}{2\rho}} \quad \text{and} \quad Z_{\text{pore}}|_{\omega\to 0} = \sqrt{\frac{\rho R_F}{2\pi^2 r^3}}\coth\sqrt{\frac{l}{\lambda}} \tag{3.48}$$

If λ is much larger than the pore length: $\lambda \gg l$, the x in the $\coth(x)$ term will be very small, and $\coth(x)$ can be replaced by $1/x$, i.e., by $(\lambda/l)^{1/2}$, which is just the same as $1/\Lambda^{1/2}1$. You should now be able to show that

$$Z_{\text{pore}}|_{\omega\to 0,\lambda\gg l} = \frac{R_\Omega}{\Lambda} = \frac{1}{2\pi rl}R_F \tag{3.49}$$

So, what is the internal area of the pore? It is the circumference times the length, i.e., exactly the $2\pi rl$ in equation (3.48). We just discovered that *the impedance is just the Faradaic impedance in* $\Omega\,cm^2$ *divided by the internal surface area of the pore*. This means we are 'seeing' the whole pore – right to the bottom. It is time to reveal that our λ *is the penetration depth* of the ac signal into the pore.

Now a task for you. Show that when λ is much less than l, the pore impedance is given by

$$Z_{\text{pore}}|_{\omega\to 0,\lambda\ll l} = \frac{R_\Omega}{\Lambda^{1/2}} = \left(\frac{\rho}{2\pi^2 r^3}\right)^{1/2}R_F^{1/2} \tag{3.50}$$

In this case, the signal does not reach the bottom of the pore.

What does ZView® have to offer for the modelling of the finite pore? We find the de Levie pore model element L_s defined as

$$Z = \left(\frac{R}{\Lambda^{\frac{1}{2}}}\right) \coth\left(\Lambda^{\frac{1}{2}}\right) \quad \text{where} \quad \Lambda = \frac{1}{A} + B(j\omega)^\phi \quad (3.51)$$

R is obviously the same as R_Ω in equation (3.42). What about Λ? This looks more complicated. Equation (3.44) defined $\Lambda = \frac{\frac{2\rho l^2}{r}}{Z_{int}}$, which is simply a constant, $\left(\frac{2\rho l^2}{r}\right)$, multiplied by $1/Z_{int}$. We will suppose that the interfacial impedance consists of the double-layer capacitance in parallel with a Faradaic resistance. Since we have a parallel RC network, the admittance $Y_{int} = 1/Z_{int}$ is given by

$$Y_{int} = \frac{1}{Z_{int}} = \frac{1}{R_F} + j\omega C_{dl} \quad (3.52)$$

You should be able to work out that $A = \left(\frac{r}{2\rho l^2}\right) R_F$, $B = \left(\frac{2\rho l^2}{r}\right) C_{dl}$ and $\phi = 1$ in equation (3.51). So, now can we model an array of identical pores with a given radius and length knowing how many pores there are per cm^2 of electrode? All we need to do is remember that admittances in parallel add. Figure 3.21 shows the result of running the de Levie simulation for a *single* pore with the dimensions shown.

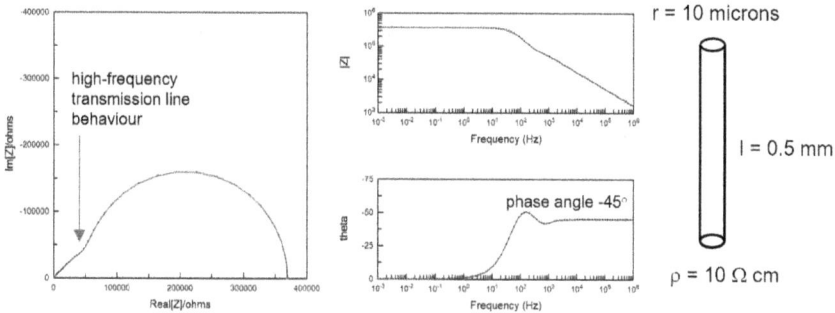

Figure 3.21. ZView® simulation using the de Levie model for a *single pore*. Pore radius $r = 10$ microns, pore length $l = 0.5$ mm. Solution resistivity $10\,\Omega$ cm, charge transfer resistance $R_F = 100\,\Omega\,\text{cm}^2$, double-layer capacitance $C_{dl} = 30\,\mu\text{F cm}^{-2}$. Values input to ZView®: $A = 2$, $B = 1.5 \times 10^{-3}$, $R = 1.59 \times 10^5$. To scale to 1 cm^2 for an array of identical pores, the impedance values are divided by the number of pores per cm^2.

The high-frequency part of the Nyquist plot corresponds to a transmission line with a phase angle of $-45°$, whereas at lower frequencies

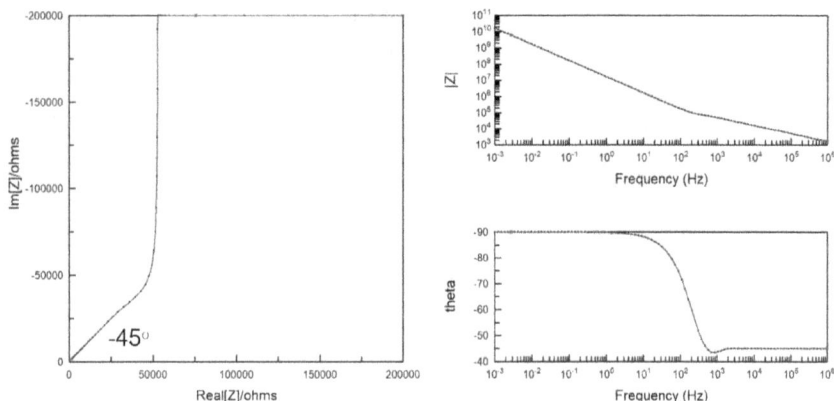

Figure 3.22. Impedance response for the single pore in the case where no charge transfer occurs. Note the high-frequency response is that of a transmission line. At lower frequencies when the ac signal penetrates to the bottom of the pore, the impedance becomes that of the double-layer capacitance of the internal surface area of the pore, so that the Nyquist plot becomes vertical as expected for a simple capacitor. Values of B and C are the same as for Figure 3.15, but B is set to 10^{10} to simulate a very high charge transfer resistance.

the impedance corresponds to an RC semicircle with a maximum at $\omega = 1/R_F C_{dl}$. $R_F(\Omega \text{ cm}^2)$ and $C_{dl}(\text{F cm}^{-2})$ can be readily obtained. We can simulate this in ZView® by using a very large value for A. Figure 3.22 shows the result obtained using the previous values of B and R and setting $A = 10^{10}$, obtained by scaling the results by the internal area of the pore, $2\pi r l$. If the charge transfer resistance is infinite (i.e., we have an *ideally polarizable* pore), the impedance response becomes that of a *transmission line*. Common examples of transmission lines are the familiar co-axial cables used for TV aerials and connections to satellite dishes. They have distributed resistance and capacitance and are designed to carry high-frequency signals with minimal power loss.

Problem

(Answer to the problem is given at the end of the book.)

3.1. Read Theory Note 3.5 again. Show that when λ is much less than l, the pore impedance is given by

$$Z_{\text{pore}}|_{\omega \to 0, \lambda \ll l} = \frac{R_\Omega}{\Lambda^{1/2}} = \left(\frac{\rho}{2\pi^2 r^3}\right)^{1/2} R_F^{1/2} \qquad (3.53)$$

In this case, the signal does not reach the bottom of the pore.

Further Reading

1. For those completely new to electrochemistry, a good starting point is a slim paperback in the Oxford Chemistry Primers series: Browne, W. R. 2018. *Electrochemistry (Oxford Chemistry Primers)*. Oxford University Press, Oxford, UK.
2. The most comprehensive reference book (over 1000 pages!) dealing with a large range of theory and experiments is Bard, A. J., Faulkner, L. R. & White, H. S. 2022. *Electrochemical Methods: Fundamentals and Applications*, 3rd edition. John Wiley & Sons, Hoboken, NJ, USA.
3. A thinner book with a lot of useful background is Compton, R. G. & Banks, C. E. 2018, *Understanding Voltammetry*, World Scientific, Singapore.
4. For the practically minded, there is Holze, R. 2019. *Experimental Electrochemistry — A Laboratory Textbook*. John Wiley & Sons, Hoboken, NJ, USA.
5. There are quite a few textbooks on impedance spectroscopy, for example, Lasia, A. 2014. *Electrochemical Impedance Spectroscopy and Its Applications*. Springer, New York.
6. Orazem, M. E. & Tribollet, B. 2017. *Electrochemical Impedance Spectroscopy*, 2nd edition. John Wiley & Sons, Hoboken, NJ, USA.
7. A recent multi-author volume that gives an idea of the breadth of the subject is Barsoukov, E. & Ross Macdonald, J. 2018. *Impedance Spectroscopy: Theory, Experiment, and Applications*, 3rd edition. John Wiley & Sons, Hoboken, NJ, USA.

Chapter 4

Kramers–Kronig Testing of Impedance Data and Inductive Loops

4.1 Introduction

Meaningful EIS measurements require that the system being studied is in a *stationary state*. This can be a problem if measurements are extended to low frequencies because the system may change during the time required to carry out the measurements. For example, just one cycle at 10 mHz takes 100 s, and to make measurements at 20 frequencies between 10 mHz and 1 Hz takes 8 min, even if only one cycle per frequency is measured. Stability on these kinds of time scales can be a problem with electrochemical systems. The question is – how can we detect whether the system is in what is called the periodic steady state? In this chapter, we take a brief look at how problems such as drift during measurements can be detected by using tests based on the *Kramers–Kronig relations* (KK for short: named after Hans Kramers and Ralph Kronig, who formulated the relations independently in 1926/7).

The KK relations connect the real and imaginary components of any 'well-behaved' complex response function. We shall not go into too much detail in this short chapter because this is quite a complicated topic. However, this should not put you off using the KK tests discussed here. You will feel a lot more confident of the validity of your impedance data if you KK test it.

During the discussion of KK tests, we will come across the idea of low-frequency *inductive loops*. These can be modelled by equivalent circuits containing inductors, but in KK fitting they can be represented by RC elements with *negative R and C values*. We will take a closer look at the possible origins of low-frequency inductive behaviour, which have been a topic of discussion for many years.

4.2   What Do We Mean by 'Drift' in the Context of Impedance Measurements?

We now look at just two of the linear circuit elements that we have encountered when describing electrochemical systems and ask why they might not remain constant as we make an EIS measurement. First, consider the Faradaic resistance, R_F. This depends on electrode potential, temperature, the local concentration of reacting species, the kinetics of the charge transfer reaction and the electrode area. R_F will therefore change if any of these parameters change. So, for example, if we slowly deplete the local concentration of reacting species, R_F will increase. Or if, during the measurement, the electrode becomes progressively blocked by contaminants from the solution, the rate of charge transfer will decrease so that R_F increases with time. Or maybe we are studying a corroding system. If corrosion progressively increases the surface area, our corrosion resistance will become smaller as time passes. In all these cases the resistance is subject to *drift*.

Second, we have the double-layer capacitance. Again, this is sensitive to adsorption of organic contaminants that block the electrode surface. We may start with a clean electrode after some cleaning procedure involving, for example, potential cycling, but during our EIS measurement, progressive contamination of the electrode by trace organic species can lead to a substantial reduction of C_{dl} as our EIS measurement progresses. C_{dl} is subject to drift.

In both examples above, the values of the circuit elements 'drift' with time. To satisfy the *periodic steady-state* condition, none of our impedance elements is permitted to change with time. If they do, we do not have a periodic steady state, and our impedance measurement will be invalid. Clearly, the longer it takes to make our measurement, the more serious the problem of drift is likely to become. Long integration times or low frequencies will compound the problem.

4.3 What Are the Kramers–Kronig Relations and Why Are They Relevant to EIS?

The KK relations crop up in different places in science, so it is worth having some idea of what they are all about, even if the theory may seem a bit impenetrable. The KK relations are widely used in optical spectroscopy to relate the real and imaginary parts of the optical dielectric constant (which are in turn related to the absorption coefficient and refractive index) of materials. In EIS, they can be used to relate the real and imaginary components of the impedance. Or to put it another way, if we know the real part of the impedance, we can calculate the imaginary part using the KK relations. Or we can convert the imaginary part to the real part. This means that the real and imaginary parts of Z that we have been dealing with all along must obey the KK relations if they are true impedances. Apologies to any mathematicians reading this – I have avoided words like 'causal', 'linear' and 'analytic in the upper complex plane'.

The KK relations define the relationships between the real and imaginary components of the impedance $Z(\omega)$ *for the fixed frequency* ω in the following way:

$$\mathrm{Re}[Z](\omega) = \frac{2}{\pi} \int_0^\infty \frac{\omega' \mathrm{Im}[Z](\omega')}{\omega^2 - \omega'^2} d\omega' \tag{4.1}$$

$$\mathrm{Im}[Z](\omega) = \frac{2}{\pi} \int_0^\infty \frac{\omega' \mathrm{Re}[Z](\omega')}{\omega^2 - \omega'^2} d\omega' \tag{4.2}$$

Note that for a *fixed frequency* ω, we are integrating with respect to $d\omega'$ and that ω' varies from zero to infinity for the integration. So, in principle, we would need to know $\mathrm{Re}(Z)$ and $\mathrm{Im}(Z)$ at all frequencies between zero and infinity. Clearly, we have a problem of 'missing bits'. We can get round this problem by some form of extrapolation to the zero and infinity limits, but this is not the place to go into any detail. Nevertheless, we can see some problems with integration straightaway. If we have a system with blocking electrodes, then we have an impedance that becomes infinite at zero frequency. In practice, of course, the impedance will be finite due to the leakage of charge across the blocking electrode, but still the impedance will be very large at zero frequency. One way round this problem is to rewrite the KK relations for admittance, because the admittance of the system with blocking electrodes will tend to zero for the frequency tending to zero. We could also have problems if we have a semi-infinite Warburg, but it turns out that this problem can also be solved.

Despite the obvious difficulty of applying the KK relations in reality, a nice way forward was formulated some time ago by Bernard Boukamp of the University of Twente (Boukamp, 1995). This is based on the fact that *any circuit made up of a series of sequential parallel RC elements obeys the KK relations.* This kind of circuit is called a Voigt circuit (named after the 19th century crystallographer Waldemar Voigt, not to be confused with the chief engineer of Messerschmitt of the same name, who in 1944 developed the first rocket-powered aeroplane, the Me163).

Using the Voigt circuit reduces the problem of testing impedance data to one of fitting the data with a sufficient number of parallel RC elements connected in series as shown in Figure 4.1. In the linear KK test, fixed RC time constants and variable resistors are used. As we shall see, R and C are permitted to take negative values.

In practice, many RC elements will be required to fit impedance plots that have features arising, for example, from finite or infinite Warburg impedances or from CPEs.

You can visit Bernard Boukamp's web site to download his KK Windows program and documentation (Boukamp). The program, which runs on Windows, allows you to load both ZPlot® and Autolab impedance files. The site also gives a description of the very useful shorthand notation developed by Boukamp for equivalent circuits.

To test the Boukamp program, I used the data calculated by ZView® for the Randles circuit in Figure 4.2. In Chapter 3, we calculated the

Figure 4.1.   Linear combination of parallel RC elements used to fit data in linear KK test. Note that some of the RC elements may have negative values.

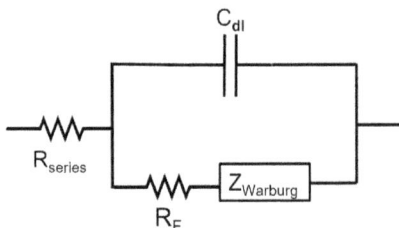

Figure 4.2.   The Randles equivalent circuit used to generate the data for the Boukamp KK test.

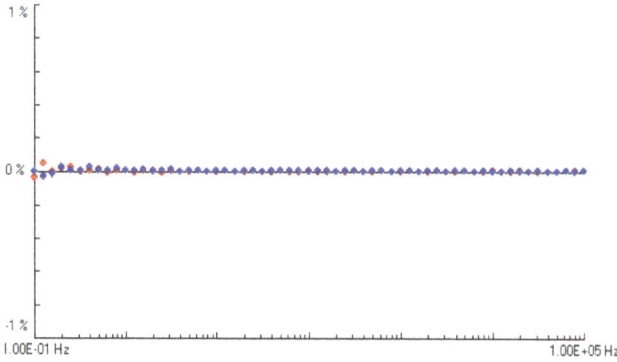

Figure 4.3.   Plot of residuals for Boukamp KK test of data simulated for the Randles circuit in Figure 4.2. Red dots: real residuals; blue dots: imaginary residuals (the red dots are largely obscured by the blue dots because the residuals are so small). The results show that the data pass the KK test as expected.

impedance of this circuit for the hexacyanoferrate system using: $\sigma = 11.9\ \Omega\,\mathrm{rad}^{1/2}\mathrm{s}^{-1/2}$, $R_F = 2.6\Omega$ and $C_{\mathrm{dl}} = 30\,\mu\mathrm{F}$. The same values were used here to calculate the impedance response with ZView®.

After loading the data file, clicking on the KK test tab results in a plot of normalized *residuals* as a function of frequency as shown in Figure 4.3. In this case we expect the data to satisfy the KK test. The residuals are shown for the real and imaginary components and are defined in each case as the difference between the measured component – $\mathrm{Re}[Z(\omega)]$ or $\mathrm{Im}[Z(\omega)]$ – and the corresponding component $\mathrm{Re}[\hat{Z}(\omega)]$ or $\mathrm{Im}[\hat{Z}(\omega)]$ calculated by KK transformation expressed as % of the magnitude of $Z(w)$.

$$\Delta\mathrm{Re}(\omega)\% = \frac{\mathrm{Re}[Z(\omega)] - \mathrm{Re}[\hat{Z}(\omega)]}{|Z(\omega)|} \times 100 \tag{4.3}$$

$$\Delta\mathrm{Im}(\omega)\% = \frac{\mathrm{Im}[Z(\omega)] - \mathrm{Im}[\hat{Z}(\omega)]}{|Z(\omega)|} \times 100 \tag{4.4}$$

Since the data were calculated using ZView®, we expect that the result should pass the KK test. It does indeed do so, as can be seen from the fact that the residuals are close to zero over the whole frequency range.

A KK test is also available in ZView®. Details of how to use it can be found in the online ZView® manual, and example data files are available to illustrate the case where the KK test fails. Another useful program called *Lin–KK Tool* Schönleber, Klotz, Ivers-Tiffée (2014), has been developed as an extension of the Boukamp approach by Michael Schönleber and his

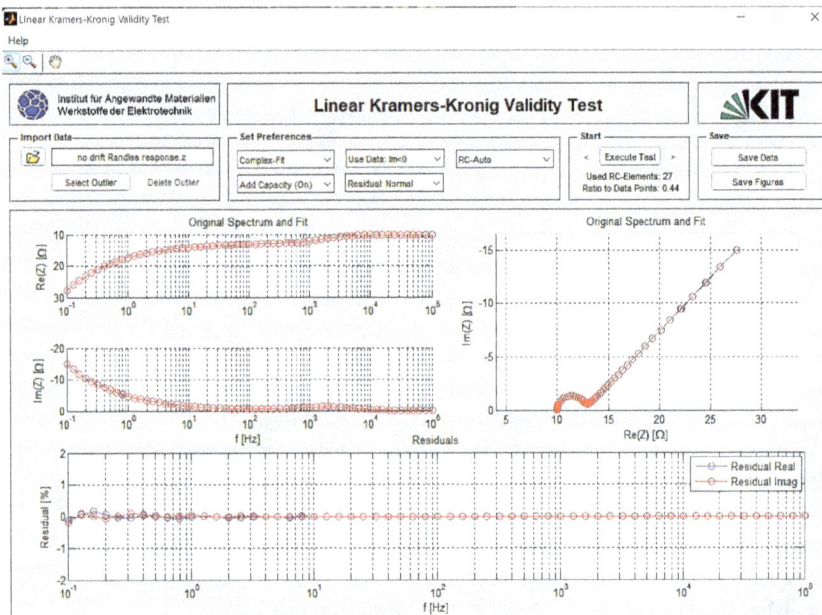

Figure 4.4. Lin–KK Test Tool Screen showing result of executing the test on an ideal Randles equivalent circuit in Figure 4.2. Note the plot of the real and imaginary residuals left after fitting. The low residual components indicate that the data passes the KK test as expected. Contrast this result with Figure 4.8 to see the larger residual components when data does not pass the KK test.

colleagues at the Karlsruhe Institute of Technology. This program also allows you to load impedance files in different formats and run the linear KK test based on the Boukamp approach.

A screen shot of the Lin–KK Tool program in action is shown in Figure 4.4. In the following section, I have used the program to see what happens to the KK test result if drift is deliberately introduced into a system modelled by the Randles equivalent circuit.

4.4 Detecting Drift Using the Lin–KK Test

To examine the effect of drift on the impedance response for the same system, let us assume that our electrode becomes progressively more contaminated as time passes and that adsorption of traces of organic compounds increases R_F and decreases C_{dl} as the surface becomes more contaminated. We will assume that the changes of R_F and C_{dl} with time are linear so that R_F has *doubled* and C_{dl} has *halved* by the time we have measured our last frequency point.

We will assume that our frequency response analyzer (FRA) has been set up to measure with an integration time of 1 s over the range 100 kHz down to 0.1 Hz, with the number of steps per decade set to 10. The measurement time will default to $1/f$ when the frequency is below 1 Hz since the minimum possible time is 1 cycle. The total time for the frequency scan is around 100 s. Figure 4.5 is a plot to show how R_F and C_{dl} are related to the elapsed time and the measurement frequency for the simulated measurement conditions.

Figure 4.6 compares the Nyquist plots calculated for the two cases – the first where R_F and C_{dl} do not change and the second for the system with drift (i.e., a non-stationary condition).

Just looking at the plot for the system with drift, one would probably not suspect that there is a problem. The general shape looks reasonable, although the slope of the Warburg part of the plot is less than the expected $-45°$ (the correct slope is shown by the data for the system without drift). The next step is to fit the data using ZView® to see if this reveals a problem. We use the equivalent circuit shown in Figure 4.2 and model the Warburg as a CPE with P allowed to vary. The result is shown in Figure 4.7.

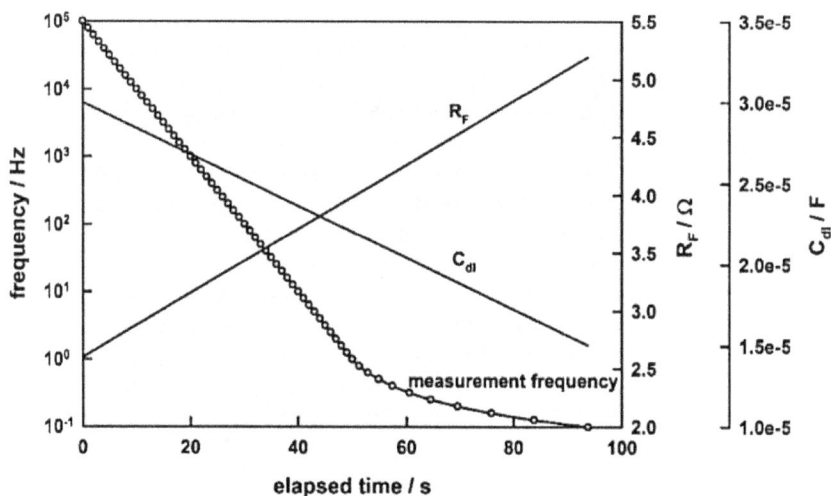

Figure 4.5. Plot showing drift of R_F and C_{dl} as a function of elapsed time during simulated EIS measurement. The plot also shows how the frequency varies with time with an integration time set to 1 s per frequency. Note that the time intervals become longer for $f < 1\,\mathrm{Hz}$, since one cycle is measured.

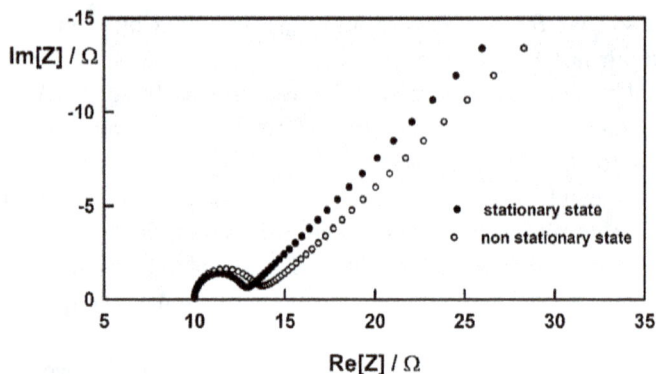

Figure 4.6. Comparison of Nyquist plots for the ideal case with no drift (stationary state) and the case where R_F increases and C_{dl} decreases with time due to electrode contamination (non-stationary state, i.e., the system is *drifting*).

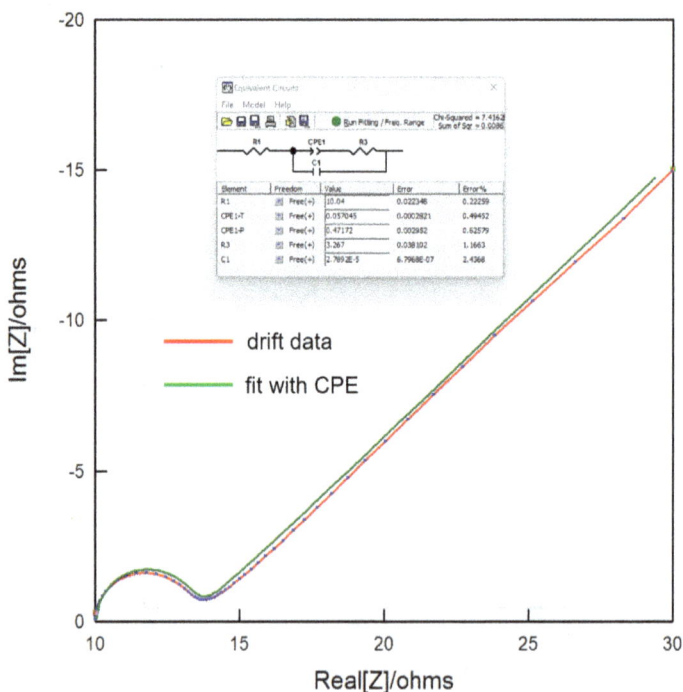

Figure 4.7. ZView® fit of data with drift. A satisfactory 'fit' is obtained using a value of P less than 0.5, but this has no obvious physical meaning. Fit values: R_{ser} 10.014 Ω, R_F 3.267 Ω, C_{dl} 27.89 μF, $T_{Warburg}$ 0.057, $P = 0.4712$ s.

Clearly, we have obtained a better fit by allowing P to vary. $P = 0.47$ does give better fit, but it is not clear what a value of P less than 0.5 means if it is supposed to be modelling a Warburg diffusion impedance. In fact, we know that the flatter Warburg slope ($P < 0.5$) is an artefact arising from the drift in R_F to higher values as the measurement progresses. The drift of R_F pushes the data points on the linear portion of the plot progressively further along to the right as we go to lower frequencies (remember we chose to measure from high to low frequencies). In this case, of course, we know the origin of the fitting problem since we simulated the 'impedance response' under non-steady-state conditions. In fact, what we have measured is *not a true impedance* since the parameter values were changing with time, i.e., we *did not have a periodic steady state*. Let us see if the Lin–KK test detects this. I have run the test with the option of including a series capacitance. This improves the fit at low frequencies. Figure 4.8 shows plots of the residuals as a function of frequency for the two data sets (the data were

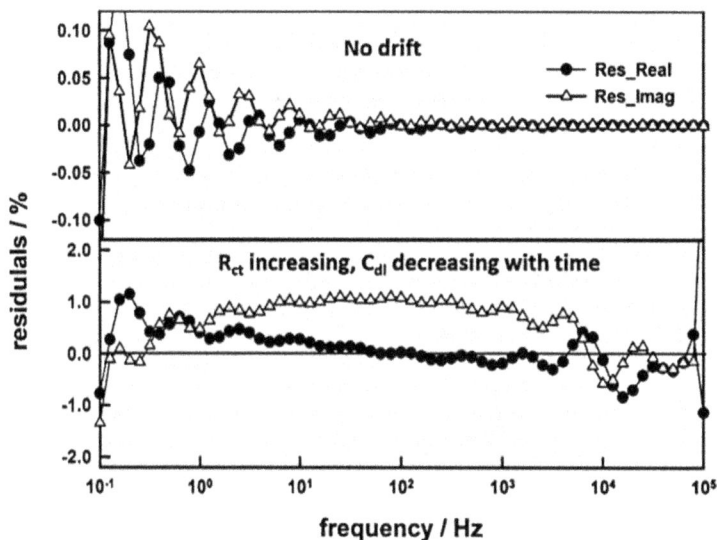

Figure 4.8. Results of the Lin–KK Tool (Schönleber) test of impedance data for the Randles circuit without drift (upper panel) and with drift in parameter values (lower panel). The 'capacitance on' option was used to obtain a better fit to the Warburg region. Note the much larger and asymmetric residual plots in the case where drift has occurred. Note also that the y scales are different in the two plots.

plotted in a separate graphics program). The first is for the true periodic steady state where all the parameter vales are constant throughout the measurement. The second is for the data with *drifting* vales of R_F and C_{dl}. In the first case, the residuals are small (<0.1%) and distributed symmetrically about zero. The oscillations at the low-frequency end arise from the fact that the Warburg is being replaced by a large number of parallel RC circuits in series – the program output tells us that 21 RC circuits were used to fit the data. In the case of the same circuit with the added drift, the residuals calculated from the test are an order of magnitude higher and not symmetrically distributed about zero. The data have *failed the Lin–KK test.*

In practice, it is a good idea to check for drift by repeating measurements under different conditions. In the previous example, the FRA had been set to measure 10 points per decade from high frequencies to low frequencies with an integration time of 1 s. If the integration time is changed to 1 cycle and the number of frequencies per decade to 7, the measurement time will be reduced substantially. Drift will then mainly affect the low-frequency data (in this case the Warburg part of the Nyquist plot) because the high-frequency data will have been measured on a much shorter time scale. Other possibilities include measuring the impedance response in the other direction, i.e., from low to high frequencies. In this case, the drift in R_{ct} will have the effect of making the Warburg plot steeper since as we go down the plot, points will be displaced progressively to the right. As a result, one would obtain different Nyquist plots depending on the direction of the frequency scan. Since failure of the KK test may indicate a nonlinear response of the electrochemical system, it is also a good idea to check that the response is independent of the amplitude of the ac signal.

We finish this chapter by applying the KK test to some real data that appear later in this book, namely the impedance of the dye-sensitized solar cell discussed in Chapter 8. In fact, we obtained a pretty good fit to our equivalent circuit, so we can be reasonably confident that the data should pass the linear KK test. The plots in Figure 4.9 confirm that this is indeed the case. The residuals show some scatter due to noise in the impedance spectrum, but the result shows that our data are KK transformable. They pass the test.

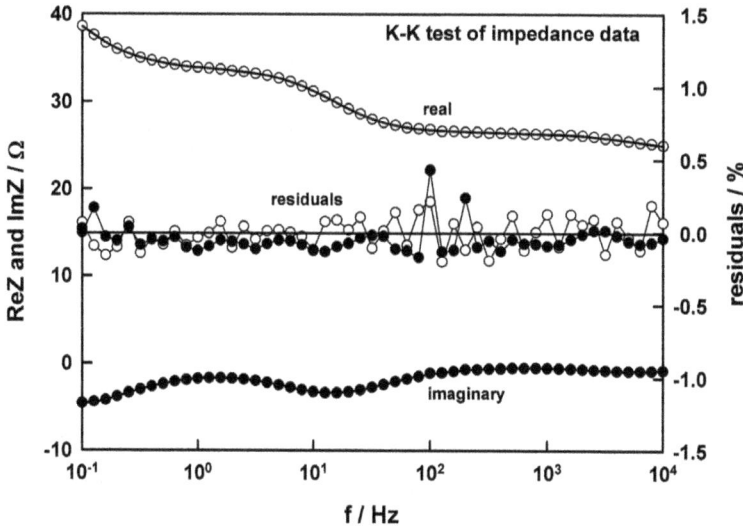

Figure 4.9. Lin–KK Tool test of the impedance data for the dye-sensitized solar cell discussed in Chapter 8. The residuals plot shows that the data set passes the KK test.

4.5 KK Testing Systems That Show a Low-Frequency Inductive Loop

Normally, we expect to see inductive effects in the high-frequency impedance behaviour of a system due to the unavoidable presence of stray inductance. However, *low-frequency inductive loops* are seen in several physical systems including fuel cells, corroding metals and solar cells. The inductance values obtained by fitting are typically so large as to be unphysical. Leaving aside the origin of the low-frequency inductive loops, this raises the interesting question of whether it is possible to carry out the KK test on systems that contain inductive elements. This problem has been addressed by Schönleber and Ivers-Tiffée (Schönleber and Ivers-Tiffee, 2015). They examined the equivalent circuit in Figure 4.10 containing resistors, capacitors and inductances and proved formally that it can be satisfactorily approximated by a Voigt circuit with a finite series of parallel RC elements with a range of time constants spaced logarithmically in time.

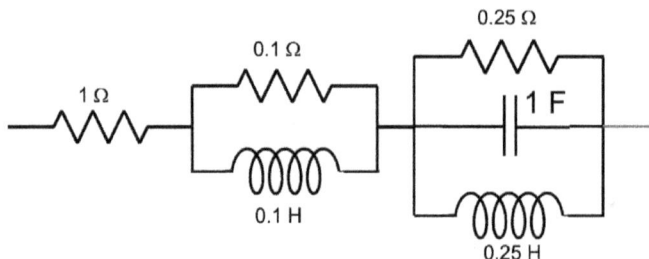

Figure 4.10. Circuit containing all three linear circuit elements used by Schönleber and Ivers-Tiffee (2015). Approximability of impedance spectra by RC elements and implications for impedance analysis. *Electrochemistry Communications*, 58, 15–19 to show formally that it can be satisfactorily approximated by a finite series of Voigt parallel *RC* elements.

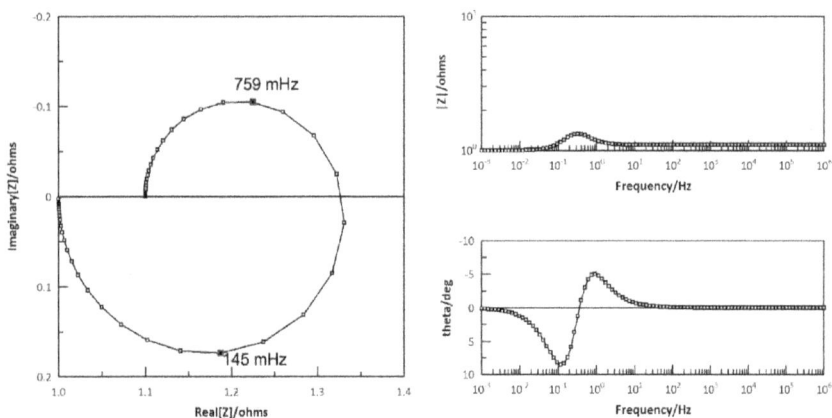

Figure 4.11. Nyquist and Bode plots of the impedance response of the circuit in Figure 4.10. Note the *RC* semicircle in the upper quadrant of the Nyquist plot and the low-frequency *LC* semicircle in the lower quadrant.

The impedance response of this circuit is illustrated in Figure 4.11.

Let us see if we can understand the Nyquist plot. At very high frequencies no current will flow through the two inductors, and the capacitance will short out the 0.25Ω parallel resistor. So, the impedance will simply be the series combination of the 1Ω and 0.1Ω resistors, i.e., 1.1Ω. At the other extreme of very low frequency, the impedance of the two inductors tends towards zero, so only the first 1Ω resistor is left. Now let us consider a radial frequency equal to the inverse of the time constant of the parallel *RC* part of the circuit. This is given by the inverse of the time *RC* constant (0.25$\Omega \times$ 1F = 0.25 s), so ω is 4 s^{-1}. What is the magnitude of the impedance of the parallel inductor at this frequency? It is $\omega L = 4 \times 0.25 = 1\,\Omega$. Since the magnitude of the impedance of the capacitor is $\frac{1}{\omega C} = 1/(4 \times 1) = 0.25\,\Omega$, most of the current will flow through the parallel RC part of the parallel circuit. So, we expect a typical RC semicircle in the upper quadrant of the Nyquist plot, which is exactly what we see. Finally, at the frequency corresponding to the minimum in the lower semicircle shown in the Nyquist plot ($2\pi \times 0.145\,\text{Hz} = 0.91\,\text{s}^{-1}$), the impedance of the capacitor is $0.91\,\text{s}^{-1} \times 1\,\text{F} = 0.91\,\Omega$, the three-element parallel circuit on the right-hand side of Figure 4.10 is approximated by a two-element parallel LC circuit, which together with the parallel LC circuit on the left-hand side gives the semicircle in the lower quadrant of the Nyquist plot.

Now that we understand the Nyquist plot, we will look at the RC circuit (Figure 4.12) that Schönleber and Ivers-Tiffée (Schönleber and Ivers-Tiffee, 2015) show mathematically gives an excellent approximation to the RLC circuit in Figure 4.11.

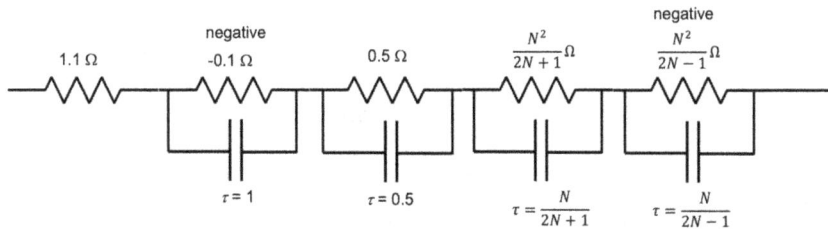

Figure 4.12. Voigt RC circuit that has an impedance that approximates the impedance of the RCL circuit in Figure 4.11. Note the circuit contains negative resistance and capacitance elements. The impedances of the two circuits approach each other more closely as N is made larger (Schönleber and Ivers-Tiffee, 2015).

There are several things to note about the circuit. First, it specifies resistances and time constants rather than resistances and capacitances. This is not a problem since the time constants are equal to the RC product, so C is easily obtained. The second is that *negative resistances* appear in the circuit. Since the time constants are always positive, it follows that the capacitances paired with negative resistances must also be negative. Finally, the last two parallel RC elements are defined in terms of a variable N. Making N larger improves the goodness of the fit as illustrated for $N = 2$ and $N = 10$ in Figure 4.13.

By now it will be clear that it is possible to fit a Voigt-type circuit with parallel RC elements to a circuit that contains inductors. This brings

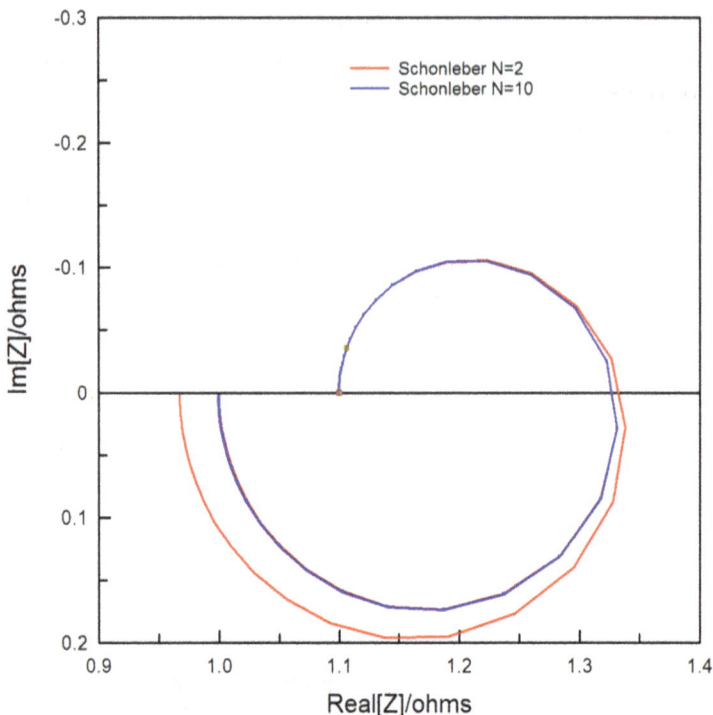

Figure 4.13. Nyquist plots comparing the impedances of the circuit in Figure 4.11 and the circuit in Figure 4.12 for $N = 2$ (red) and $N = 10$ (blue). The agreement for $N = 10$ is so good that the difference between the RCL circuit plot is not visible.

us back to the idea that *circuits may be degenerate* as we discussed in Chapter 3. We will demonstrate this by looking at two equivalent circuits that generate a loop in the lower quadrant of the Nyquist plot. The left-hand side of Figure 4.14 shows a circuit with an inductor that produces a semicircle in the upper quadrant of the Nyquist plot followed by a loop in the lower quadrant at low frequencies. The right-hand circuit gives the same response *if the values of R3 and C2 are negative.*

We can demonstrate the degeneracy by creating a Nyquist plot using the circuit with the inductor and then fitting it with the circuit with negative $R3$ and $C2$. Figure 4.15 shows that the fit is perfect. This is because the two circuits are degenerate. This explains why circuits with inductances can be simulated by an appropriate series of RC elements, provided that negative values of R and C are permitted, despite the fact that their physical meaning may not be clear.

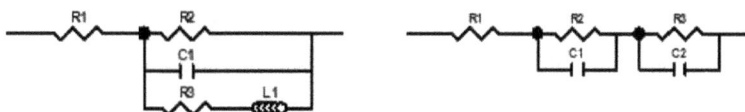

Figure 4.14. Two equivalent circuits that are degenerate. The degeneracy shows that some circuits containing inductive elements can be represented exactly by a circuit containing a finite number of resistances and capacitances.

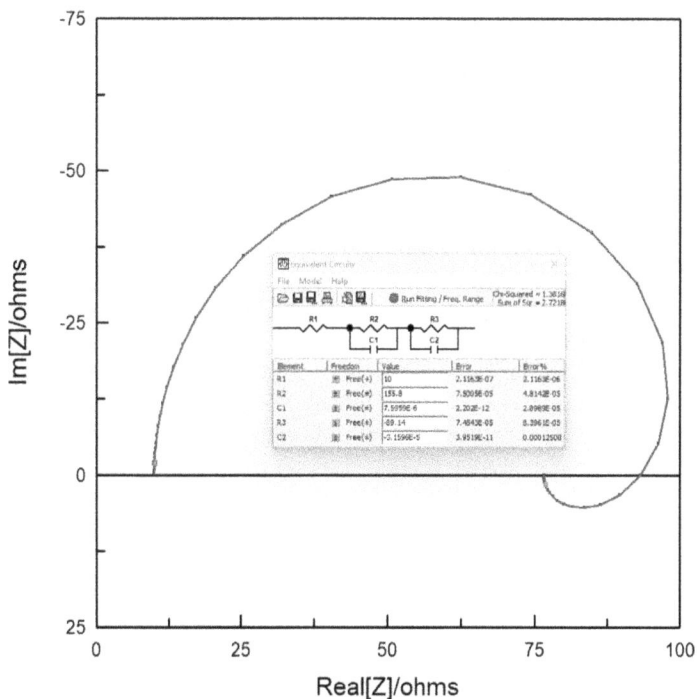

Figure 4.15. Points: simulated response for the equivalent circuit on the left-hand side of Figure 4.13. The values used in the simulation are shown in the figure. Line of best fit to the impedance response using the equivalent circuit on the right-hand side of Figure 4.3 and allowing negative values of $R3$ and $C2$. The values obtained by fitting are shown in the figure. It can be shown that the two equivalent circuits in Figure 4.3 are degenerate if *negative values of resistance and capacitance* are allowed in the right-hand circuit.

Although inductive loops have been observed for a range of different electrochemical systems and for solar cells, care is needed to ensure the loop is not an artefact of the measurement. For a full discussion of inductive behaviour, the reader is referred to a paper by Bisquert and Guerrero

(Bisquert and Guerrero, 2022) that defines the concept of a *chemical inductor* and shows how it can arise from the kinetics of electrochemical, biological and semiconductor systems. The important point as far as this chapter is concerned is that the chemical inductor behaves formally as an inductance in terms of its frequency-dependent impedance, but the origin of the inductive effect differs completely from that operating in a conventional electrical inductor.

References

Bisquert, J. & Guerrero, A. 2022. Chemical inductor. *Journal of the American Chemical Society*, 144, 5996–6009.

Boukamp, B. A. 1995. *Kramers–Kronig data validation programme (ZIP)*. University of Twente. Available: https://www.utwente.nl/en/tnw/ims/publications/downloads/ Accessed 4th July 2023.

Boukamp, B. A. 1995. A linear Kronig Kramers transform test for immittance data validation. *Journal of the Electrochemical Society*, 142, 1885–1894.

Schönleber, M. & Ivers-Tiffee, E. 2015. Approximability of impedance spectra by RC elements and implications for impedance analysis. *Electrochemistry Communications*, 58, 15–19.

Schönleber, M., Klotz, D. & Ivers-Tiffée, E. 2014. *Kramers–Kronig Validity Test Lin-KK for Impedance Spectra* [Online]. Karlsruhe Institute of Technology. Available: https://www.iam.kit.edu/et/english/Lin-KK.php Accessed 4th July 2023.

Schönleber, M., Klotz, D. & Ivers-Tiffée, E. 2014. A method for improving the robustness of linear Kramers-Kronig validity tests. *Electrochimica Acta*, 131, 20–27.

Chapter 5

The Potentiostat and the Frequency-Response Analyzer: How Do They Work?

5.1 Introduction

It is easy to be seduced by the simplicity of carrying out impedance measurements with modern computer-controlled equipment. The results appear in real time on the screen, and fitting is straightforward using commercial software. However, knowledge of what is going on 'behind the scenes' is essential. The potentiostat and the frequency-response analyzer (FRA) are critical parts of any modern impedance measurement, and this chapter aims to explain how they work. We also take a quick look at the digital signal processing that takes place in the FRA to produce the familiar Nyquist and Bode plots. The chapter concludes with an explanation of the Fourier Transform (FT) and explains how the discrete fast Fourier transform (FFT) is used to analyze signals containing simultaneous sinusoidal components with different frequencies in the multisine method for impedance measurements.

5.2 The Ideal Potentiostat

The potentiostat is a *control device* that is based on the concept of *negative feedback*. Negative feedback is a stabilizing process in which any deviation from a pre-defined value is fed back in such a way that the system is driven back towards the desired state. The modern world is full of negative feedback mechanisms to ensure stability. Your body temperature is carefully regulated by negative feedback. If you are too hot, your sweat glands produce sweat that evaporates and cools your body. If you are cold, your muscles contract rapidly as you shiver, and the heat generated tends to increase your temperature back towards its normal value. Modern aeroplanes use an autopilot which detects any deviations from a given three-dimensional position and speed and alters the controls in such a way as to reduce the deviations to zero. The job of the potentiostat is to detect any deviations of the electrode potential from a given value and to push the potential back to its correct value.

At the heart of the potentiostat is a device called an *operational amplifier*. The term 'operational' comes from the early uses of electronic amplifiers to carry out mathematical operations such as addition, subtraction, differentiation and integration in an *analogue computer*. Unlike a digital computer, which works on binary information, i.e., 0 and 1, an analogue computer processes continuously varying signals using *negative feedback loops*.

To understand how operational amplifiers and negative feedback loops work, we start by using the hypothetical idea of an *ideal operational amplifier* (OA). This amplifier is a three-terminal device shown by the triangle symbol in Figure 5.1. It has two inputs – labelled positive and negative on the figure – and one output. The positive input is called the *non-inverting input* and the negative input is the *inverting input*.

The OA is a differential amplifier – it amplifies the voltage difference between the input terminals by an amount called the open loop gain, G. Real operational amplifiers have very high open loop gains of 10^6 or more for dc signals, so we will assume that our ideal OA has an open loop gain of infinity. The input resistance of real OAs can be very high – up to $10^{12}\ \Omega$, so we let our ideal OA have an input resistance of infinity. Finally, the output resistance of real OAs is very low – typically a few 10s of ohms, so we will let our ideal OA have an output resistance of zero.

So, is our ideal OA a useful device? At first sight, the answer appears to be no. Any finite voltage difference between the inputs will result in a very

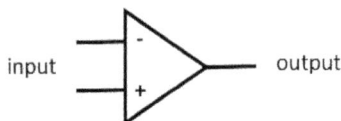

Figure 5.1. The operational amplifier (OA). The hypothetical 'ideal' OA has infinite open loop gain at all frequencies, an infinite input resistance and zero output resistance.

negative feedback loop

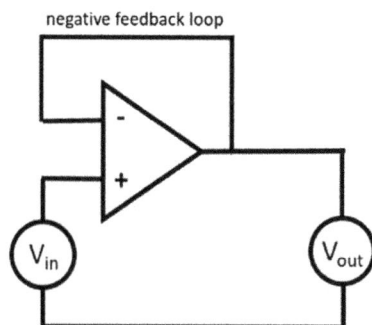

Figure 5.2. The negative feedback loop used to construct a voltage follower using an operational amplifier.

uncomfortable infinite output voltage. So how do we tame the monster we have created? The answer is, of course, negative feedback. We can connect the output terminal back to the negative input terminal and connect a voltage source to the negative input as shown in Figure 5.2 to create a device called a *voltage follower*.

What are the properties of this device? Any difference in voltage between the voltages going to the positive and negative input terminals will be amplified and fed back to the negative input terminal. Since the gain of our ideal OA is infinite, the difference between the inputs will be driven to zero. This means that the output voltage will be forced to be the same as the input voltage: the output 'follows' the input – hence the name. This does not look particularly useful at first sight. Why not just use the input voltage device on its own? All voltage sources have a finite output resistance R_{out}, which means that as we draw current from them, the voltage will decrease by an amount $i_{out} \times R_{out}$. The nice thing about our ideal OA is that it does not draw any current from the voltage source because the input impedance of the OA is infinite. However, the output impedance of our voltage follower is zero, so we no longer have the problem of the voltage dropping when we draw current. This means we can use a voltage follower to connect to a reference electrode in an electrochemical cell without drawing currents that would disturb its equilibrium potential. We have the first part of our potentiostat.

A potentiostat not only has to control the potential of an electrode with respect to a reference electrode, it also needs to measure the current. We could do this by inserting a resistance in the circuit and measuring the voltage drop across the resistance with a differential amplifier, but, ideally, we would like to measure the current without adding a resistance to the circuit. This is another task that can be performed using an operational amplifier with a negative feedback loop. In this case, the feedback loop contains a resistance, but as far as the electrochemical cell is concerned, it looks as if the current is simply flowing directly to ground with no resistor in the pathway. The OA device that achieves this trick is the *current follower* (or *zero-resistance ammeter*, as it is sometimes called). The current follower circuit is shown in Figure 5.3.

Note that the positive terminal of the OA is connected to the circuit common or, in this case, ground. The negative feedback loop ensures that the negative terminal is also effectively at ground – we say it is a *virtual ground*. So, from the point of view of any current source, it looks as if the current is flowing to ground, i.e., there is no resistor in the way. But of course, the current must go somewhere. Since the input resistance of the OA is infinite, the only place it can go is through the negative feedback loop. In doing so, it must pass through the feedback resistor R and the result is that the output voltage is $i_{input} \times R$. As Figure 5.3 shows, the current actually flows into the output of the amplifier, or in other words, the OA generates exactly the right current of opposite sign to ensure that the negative input terminal remains at virtual ground. This makes sense if you look back at Section 2.3 where we discussed Kirchhoff's current law. You will recognize that the virtual ground point is a *node*, and Kirchhoff

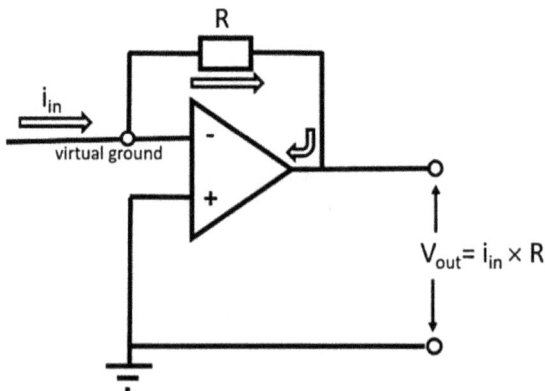

Figure 5.3. The current follower or zero-resistance ammeter. The negative feedback loop ensures that the negative input terminal of the OA is held at *virtual ground*, i.e., the voltage difference between ground and the negative input is forced to zero. The output voltage is a direct measure of the input current.

reminds us that the algebraic sum of the currents flowing into any node must be zero. So, the current in the feedback loop is given a minus sign. The result is that we have an output voltage that is a direct measure of the current we want to measure, but we did not 'load' the circuit with a resistor between the current source and (virtual) ground. Neat – we have the second component of our potentiostat.

When carrying out impedance measurements we would like to have independent control of the dc potential and of the ac signal. We might, for example, want to scan or step the potential with a superimposed ac signal. Therefore, what we need is some way of adding various input signals. By now it will come as no surprise to learn that our versatile friend the OA can come to our rescue. What we need is an *adder*. Figure 5.4 shows how we can construct an adder, again making use of the virtual ground regulated by the negative feedback loop. In this example, we have three voltage sources, each connected to virtual ground by an identical resistor. The currents flowing into virtual ground are then V_1/R, V_2/R and V_3/R. The *sum* of all three currents then flows through the feedback resistor, which has the same value as the input resistors. I shall leave it to you to work out that the output voltage must be the sum of the input voltages, $V_1 + V_2 + V_3$ (remember that the virtual ground point at the inverting input of the OA is a node, so you need to apply Kirchhoff's current law). We now have the adder we need to create complicated signals to apply to our system.

This leaves us with just the control of the potential of the working electrode in the cell to consider. Again, we can use our ideal OA with

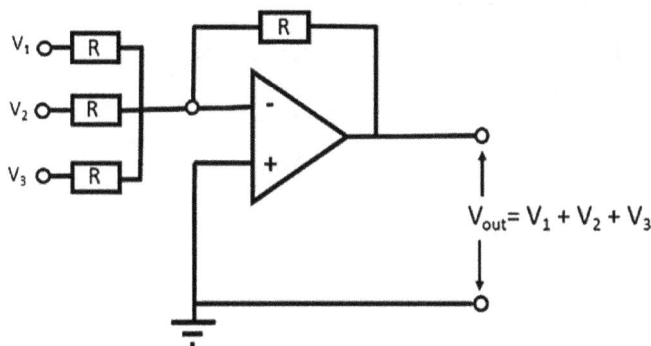

Figure 5.4. The adder. The device sums the currents V_1/R, V_2/R and V_3/R flowing into the virtual ground at the negative terminal. The current flowing through the feedback resistor is $-(V_1/R + V_2/R + V_3/R)$. The voltage at the output is therefore $R(V_1/R + V_2/R + V_3/R) = V_1 + V_2 + V_3$.

negative feedback loop to achieve this. The obvious starting point is to connect our working electrode to a current follower so we can measure the current and have the working electrode at virtual ground. Now we need to sense the difference in potential between the working and reference electrodes and ensure that this corresponds to our input voltage (or sum of input voltages). But we must be careful not to pass any current through the reference electrode. Passing the current is the job of the secondary (or counter) electrode in our *three-electrode cell.* Figure 5.5 shows how we can do this. If you work your way round the circuit remembering to apply Kirchhoff's laws, you should be able to confirm that the potentiostat does indeed control the potential difference between working and reference electrode at the desired values (in this case, the sum of the three input voltages). There are hints in Figure 5.5 to help you.

The potentiostat controls the working electrode potential without passing any current through the reference electrode, which can therefore happily stay at its equilibrium potential. All the current through the cell flows between the secondary and working electrodes. Note that we can measure the actual potential $R_{\mathrm{RE}} - E_{\mathrm{WE}}$ at the output terminal of the voltage follower. As we shall see later, this is important for impedance measurements.

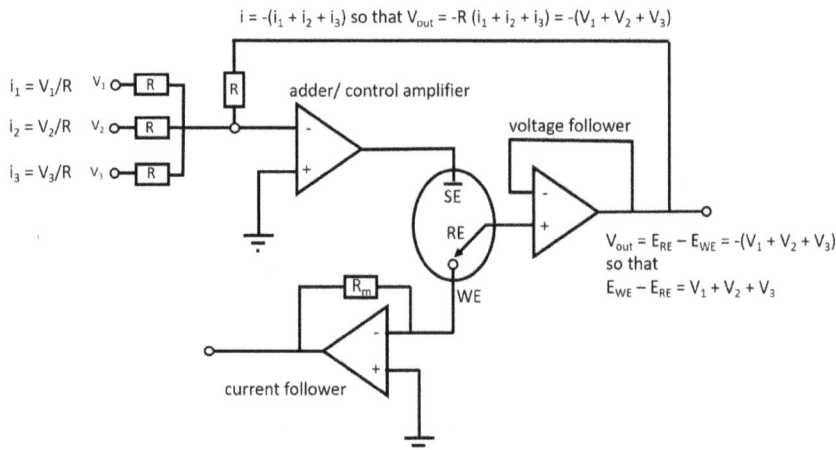

$$i = -(i_1 + i_2 + i_3) \text{ so that } V_{\mathrm{out}} = -R(i_1 + i_2 + i_3) = -(V_1 + V_2 + V_3)$$

$i_1 = V_1/R$

$i_2 = V_2/R$

$i_3 = V_3/R$

adder/ control amplifier

voltage follower

$V_{\mathrm{out}} = E_{\mathrm{RE}} - E_{\mathrm{WE}} = -(V_1 + V_2 + V_3)$
so that
$E_{\mathrm{WE}} - E_{\mathrm{RE}} = V_1 + V_2 + V_3$

current follower

Figure 5.5. A simple potentiostat circuit using three operational amplifiers. Work your way around the circuit using Kirchhoff's laws (see Chapter 2) to see if you agree that it will do the job.

5.3 The Real Potentiostat

You can see that negative feedback leads to stability and control. But what happens if while the signal is passing around the feedback loop it changes phase by 180° due to effects such as inductance, stray capacitance and amplifier characteristics? Bad news. Our feedback is now *positive*. This is what happens with a badly set up audio system in a lecture room. As the lecturer with the microphone approaches the loudspeaker, an excruciating howling noise builds up rapidly. This is due to positive feedback from the loudspeaker to the microphone and via the audio amplifier back to the loudspeaker. We certainly do not want this to happen with any OA control circuit. Since the phase shift increases with increasing frequency, the designers of OAs must ensure that when the phase shift reaches 180° at high frequencies, the amplifier gain is less than 1. This avoids positive feedback, but it does so at the cost of abandoning one of the characteristics of our ideal OA – the infinite gain (at all frequencies). Instead, the open loop gain is tailored to decrease linearly with increasing frequency until at a particular frequency it passes through unity (typically somewhere in the MHz frequency range).

Operational amplifiers are a topic in themselves, and we won't go into any further details here. If you are interested in finding out more, a good starting point is the Analog Devices website (Analog_Devices). All we do need to note at this point is that real potentiostats are going to run into problems at high frequencies. This means that the actual sinusoidal modulation of the electrode potential will not be the same as the input sinewave signal – it will *be attenuated and phase shifted* relative to the input signal. This is why when defining the impedance as the complex ratio of voltage to current we need to measure the *actual ac potential difference* between the reference and working electrodes (which may be attenuated and phase shifted relative to the ac programming signal) as well as the current through the cell. If we use the input ac voltage to obtain the ratio $\frac{V}{I} = Z$, we are going to get the wrong answer at high frequencies. The potentiostats used for EIS measurements are designed to avoid this problem.

Modern instrumentation for EIS allows measurements to be made routinely up to 1 MHz, with possible extension to 10 MHz with specialized equipment. However, the accuracy of the measurement depends on the magnitude of the impedance being measured. The 1 MHz limit generally applies only to impedances in the range $10–10^3\,\Omega$. To compare the

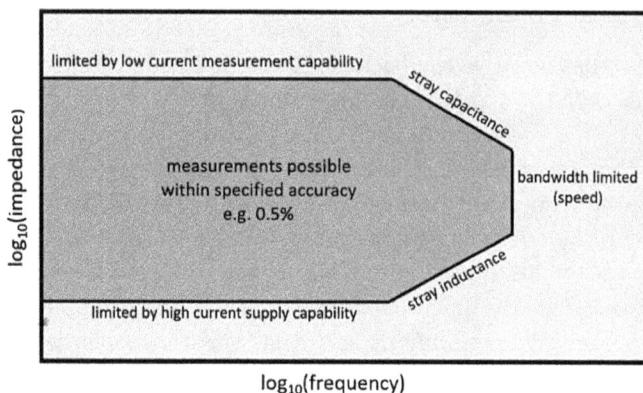

Figure 5.6. General shape of an impedance accuracy contour plot specifying the 'safe' region for impedance measurements. Most instrument manufacturers provide this information on their website, and it is good practice to use the plot to check that you can obtain an acceptable accuracy for your measurements.

performance of different potentiostat/FRA systems, look for the *impedance accuracy contour plot* in the manufacturers' specification. These plots define the two-dimensional impedance/frequency region in which reliable EIS measurements can be made within a specified accuracy. These plots have the general form shown in Figure 5.6. The speed with which the potentiostat can respond is obviously important (this determines the *bandwidth*). But stray capacitance and stray inductance can never be reduced to zero, and these affect the accuracy of measurements as shown in the contour plot. Low impedances mean high currents, and this means more problems with inductance. High impedances mean that it is easy to bypass the impedance through even very small stray capacitances. Keeping cables short and thinking about the positioning of cell connections can help to reduce the stray L and C effects. High-resistance reference electrodes with porous frits can be an additional problem, particularly if the frit becomes partially blocked, because they give rise to additional phase shift and decrease potentiostat stability. If your reference electrode has an impedance of greater than $10^3\,\Omega$, you will have problems with high-frequency measurements. For a discussion on an interesting way of minimizing reference electrode effects, see Tran *et al.* (2011).

The commercial potentiostats used for impedance measurements are more sophisticated than the basic design shown in Figure 5.5, but their working principle is the same. In most cases, specialized potentiostats have four terminals so that the potential difference across the cell can be measured directly between the working and reference electrodes rather than between the reference electrode and virtual ground. This is important for high-current applications where there may be a significant potential drop in the lead connecting the working electrode to the potentiostat. The key requirement is that the potentiostat measures the potential difference between the reference and working electrodes and the current through the cell with high accuracy when operating in the 'safe' zone shown in the impedance accuracy contour plot. The four-terminal configuration can also be used to study membranes, as shown in Figure 5.7. In this case, it is the potential difference across the membrane that is controlled rather than the potential of a working electrode.

Most commercial potentiostats can also operate in *galvanostatic mode*. This means that the controlled quantity is the current rather than the voltage. In this case, it is necessary to ensure that the voltage produced by the ac current remains small enough for the linear approximation to be valid. The upper limit of the accessible frequency range is generally lower for galvanostatic operation than for potentiostatic operation.

Normally, as a user you will be looking at Nyquist or Bode plots of impedance, i.e., at a ratio of voltage to current. However, it is good practice to look at the individual (single channel) voltage and current signals separately if you are experiencing problems. This will show the attenuation and phase shift in the voltage signal as well as reveal if there are problems with the magnitude of the current. If the measured impedance is very high, the current will be correspondingly small and the signal/noise ratio, high. For measurements of very high or very low impedances, you can find helpful documentation and advice on manufacturers websites.

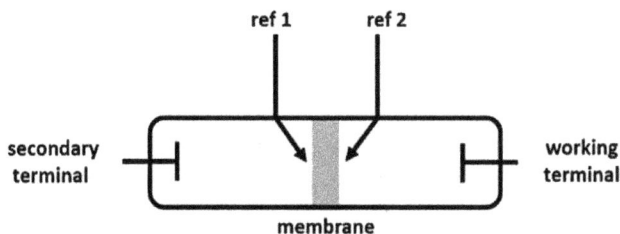

Figure 5.7. The four-terminal potentiostat configuration used to measure the impedance of membranes. Two reference electrodes are used to control the ac potential across the membrane and the current flows through the membrane via the other two electrodes.

5.4 Signal Processing in the FRA: Single Sine Correlation – The Ideal Linear Case

Now it is time to turn our attention to the frequency-response analyzer. The FRA uses *digital signal processing*: the sinusoidal voltage is generated digitally, and the response is also converted to digital form to allow it to be manipulated as required. How does it work out the real and imaginary components of the impedance? It can do this by a technique called *single sine correlation*. Essentially, this involves multiplying the response signal by either a reference sine wave or a cosine wave and integrating the result as shown in the schematical representation in Figure 5.8.

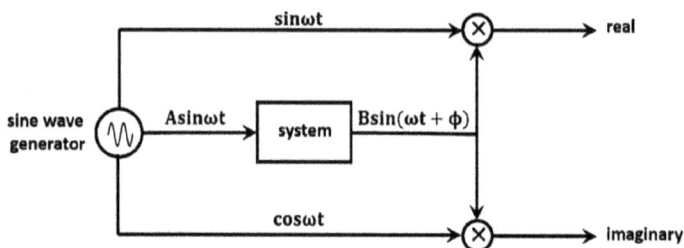

Figure 5.8. Single sine correlation. The FRA imposes a sinusoidal voltage modulation on the system. The resulting current response is multiplied by the reference sine function to obtain the real component and by the reference cosine function to obtain the imaginary component.

To get an idea of how this works, we begin by assuming that our system response is linear and has no superimposed noise. To start, we will assume that we are looking at the impedance of a simple resistor, so that the current will be in phase with the voltage as shown in the top half of Figure 5.9. The bottom half of the figure shows what happens when we multiply the current signal by the reference sine wave and by the corresponding cosine wave.

Note that multiplication by the sine wave gives a response that corresponds to the *square* of a sine wave. It has double the frequency, and it is *entirely positive*. So, if we *integrate* this signal, we will get a positive value (integration corresponds to finding the area under the plot, taking into account the positive and negative signs of the signal). Now look at what happened when we multiplied the current signal by the cosine reference signal. Again, we get a signal with twice the frequency, but now it is symmetrical about the zero axis. If we integrate this signal, the positive (up) and negative (down) areas cancel out to give zero. You can see that

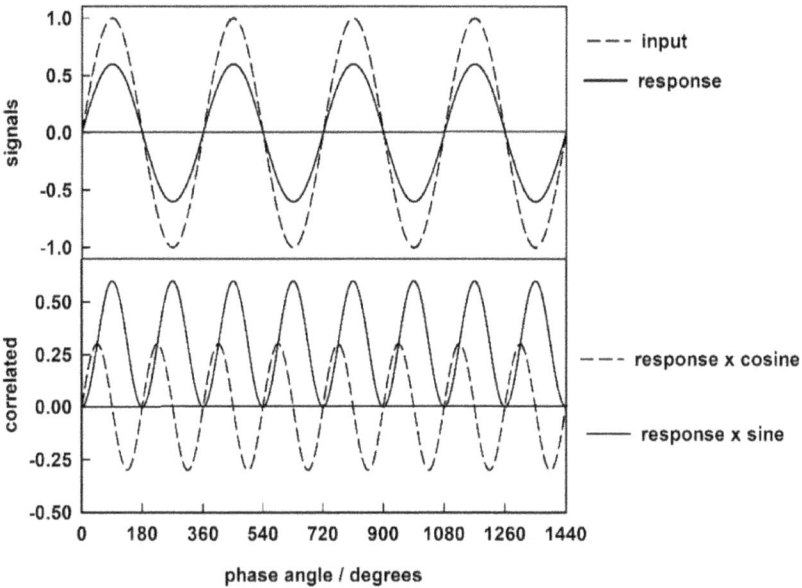

Figure 5.9. Single sine correlation for the case where the current is in phase with the voltage (the impedance is a resistor). The figure shows what happens when we multiply the response signal by sine and cosine waves. Integration of the response × cosine function will give a positive value, whereas integration of the signal × cosine function will give zero. Note that integration corresponds to finding the *area* under the plot, considering the positive (up) and negative (down) signs of the signal.

the signal obtained by the sine multiplication is the in-phase component of the signal, i.e., it corresponds to the real value of the impedance. Cosine multiplication and integration give the 90° phase-shifted value of the impedance, which in this case is zero. This is what we expect since the impedance of a resistor is real.

You should be able to see that if our impedance was a capacitor so that the current is 90° out of phase with the voltage, it will now be the response × sine signal that integrates to zero and the response × cosine signal that integrates to give a finite value. In other words, the impedance will be imaginary.

Now let us suppose that our impedance is a semi-infinite Warburg with a phase shift of −45°. Figure 5.10 shows the result of the correlation in this case. Again, the upper part of the figure shows the voltage input and the current response, and the lower part shows the results obtained by single sine correlation.

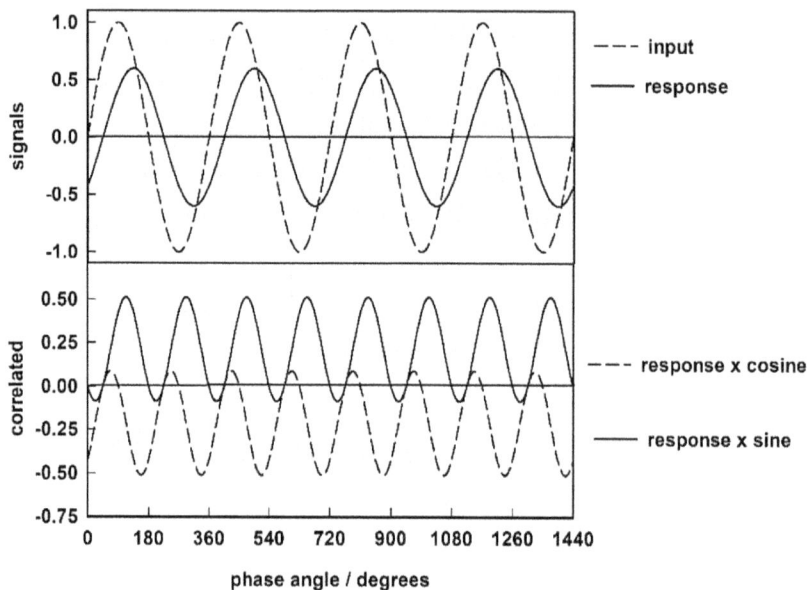

Figure 5.10. Result of single sine correlation for the case of a response that is shifted by 45° from the input. The results of integration of the real and imaginary components are equal but opposite in sign. The real component is positive, and the imaginary component is negative.

In the preceding examples, it does not matter whether we integrate just one full cycle or whether we integrate several and divide by the number of cycles – we will get the same result.

But what happens if our signal is not nice and clean? Instead, we have some superimposed random noise or pickup at mains frequency (or multiples of mains frequency) due to nearby equipment. How will this affect the result? Or maybe we have a nonlinear response because the amplitude of our input signal is too large to ensure a linear response. This is where the integration of the correlated signal by the FRA is very useful. The integration improves the accuracy of the measurement but slows down the frequency scan.

To explore the effect of integrating more than one cycle, I have generated a 'noisy' current response with a frequency of 1 kHz and a phase shift of −45° (Warburg behaviour) using a random number generator to add the noise to the current response. The response signal is shown at the top of Figure 5.11. You can still see the sinusoidal response, even though the noise amplitude is even larger than the signal we want to analyze using the

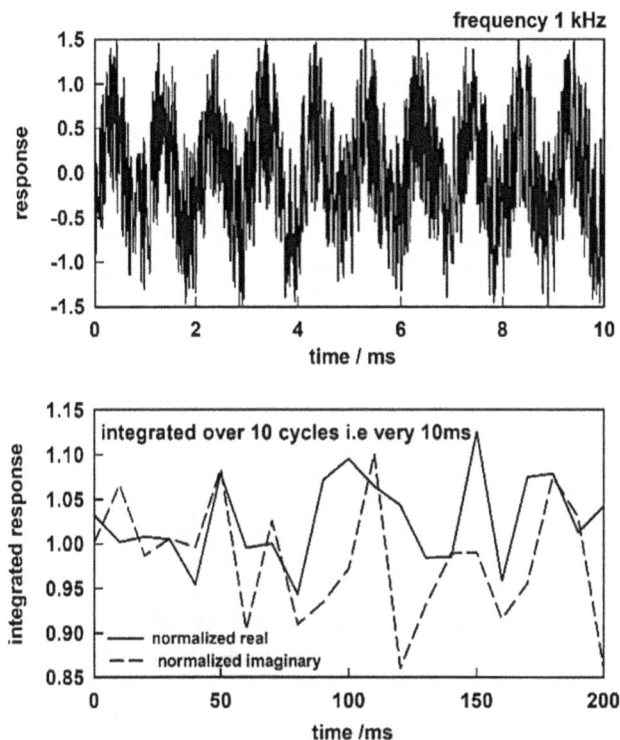

Figure 5.11. The top figure shows the noisy current signal for a Warburg impedance (the current is phase shifted by +45° relative to the voltage, which is not shown here). The lower figure shows the result of repeatedly integrating and averaging (20 times over 10 cycles) the real and imaginary components of the response obtained by correlation, i.e., by multiplying the response by the reference sine and cosine, respectively. Since the frequency is 1 kHz, the integration requires 10 ms. The error is around ± 10%. A tenfold improvement in signal to noise ratio can be obtained by integrating 100 times longer, i.e., for 1 s. This will reduce the error below 1% at the cost of slowing down the measurement.

correlation technique. So, how do we do the averaging? We simply integrate the response signal over a fixed number of cycles and divide the answer by the number of cycles. This corresponds to averaging the signal. If we are dealing with *random noise*, the integral of the noise should become smaller the longer we integrate since the chances of positive or negative deviations from the true signal are the same.

I have chosen to integrate 10 cycles in this example and have repeated the integration 20 times using a newly generated noisy waveform each time to show how even such a short integration reduces the errors in the real and imaginary components to around ± 10% (in the figure the real and imaginary components have been normalized by dividing by their true values – remember that the imaginary part of the Warburg is negative). The lower part of the figure shows the result of repeating the integration of 10 cycles 20 times. Since each sample of 10 cycles lasts 10 ms, the total time taken is 200 ms. You can see that the results for the real and imaginary components fluctuate above and below the correct value (1.0) by up to 10%. In fact, the errors decrease with the $1/\sqrt{N}$, where N is the number of cycles sampled, so that a reduction of errors below 1% would require 1000 cycles (this would correspond to 1 s since the frequency is 1 kHz). The takeaway message here is that longer averaging will reduce 'noise' in the impedance output data, but the downside is that it slows down the measurement. Averaging 1000 cycles is not too bad for 1 kHz, but for 1 Hz, it would take 17 min – far too long for practical applications. For this reason, most FRAs allow you to choose an integration time as an alternative to the number of cycles. So, if you choose 1 s, the FRA will integrate 1000 cycles at 1 kHz but only one cycle at 1 Hz. For still lower frequencies, the FRA will default to integrating just one cycle.

Now we will look at another common problem – interference. Mains frequency interference often occurs due to pickup from nearby equipment. What this means is that the signal that the FRA is trying to analyze is the sum of the true response of your system plus the mains pickup. The FRA is very good at rejecting frequencies other than the measurement frequency because the correlation technique multiplies the response signal by the reference sine and cosine signals. Of course, if your measurement frequency is equal to the mains frequency, then the FRA will run into problems. For this reason, it is a good idea to avoid mains frequencies (and higher harmonics of the mains frequency) when setting up your frequency scan. The start and finish frequencies as well as the number of steps per decade can be varied to avoid frequencies that are harmonics of the mains frequency.

To test out how good the correlation technique is at rejecting frequencies other than the measurement, I have chosen a measurement frequency of 30 Hz and have superimposed on it a 50 Hz signal of equal amplitude and random phase. The phase shift is set to $-45°$ again for a Warburg impedance. The response signal looks alarming – you can see the 50 Hz signal completely distorts the waveform that the FRA has to analyze. So, we can again repeatedly integrate 10 cycles of our poor-quality signal to see what we get for the real and imaginary components.

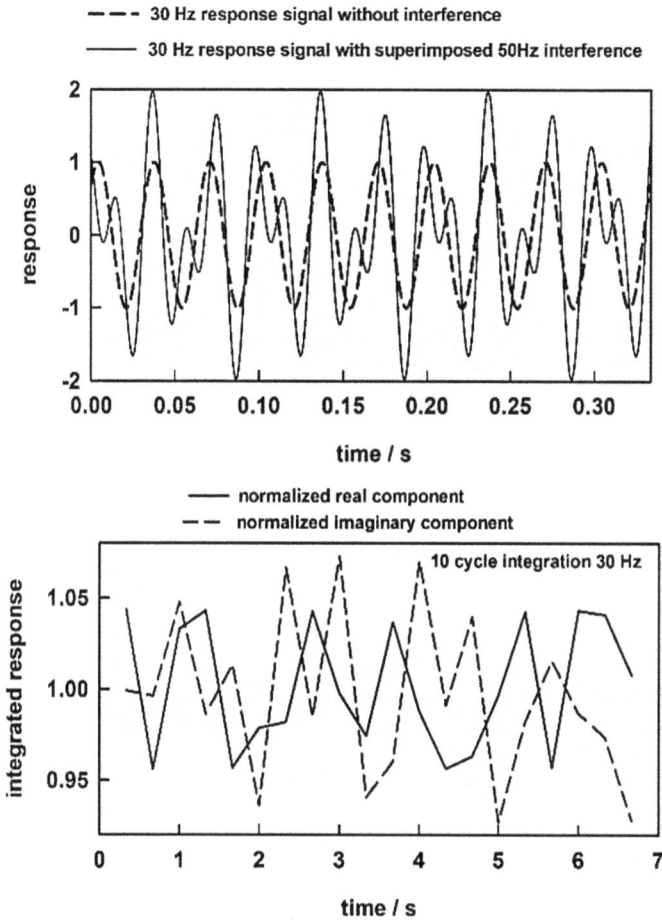

Figure 5.12. Single sine correlation of a 30 Hz response signal with superimposed 50 Hz interference. Repeated integration (20 times over 10 cycles, taking 1/3 of a second each time) reduces the errors to around ± 5%.

Figure 5.12 shows our signal and the results of the correlation and integration.

There is much more to the correlation story, and for further information, I recommend that you look at some of the information available online. For example, Figure 5.13, taken from a Solartron Technical Report (Cogger and Evans, 1999), shows rejection curves as a function of the number of cycles integrated. I suggest you explore this topic if you are interested.

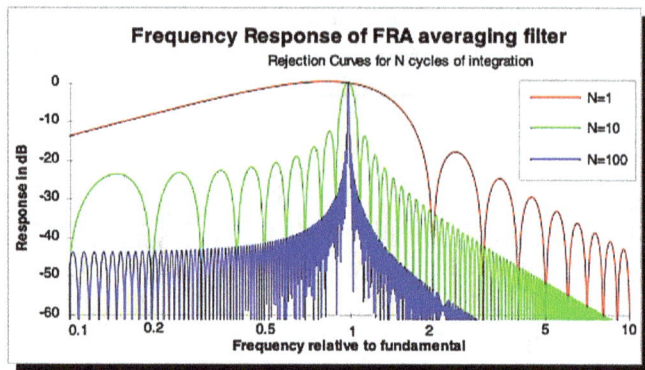

Figure 5.13. Frequency response of the averaging filter in a frequency-response analyzer reproduced with permission from Cogger and Evans (1999).

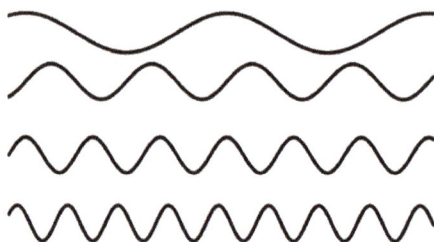

Figure 5.14. Multisine methods. Spaghetti salad: different frequency signals with different phases added to produce a voltage perturbation signal. We need to 'unscramble' the current response to find the real and imaginary components of the impedance at each of the frequencies. This is done by using an FFT algorithm.

5.5 Multisine Fourier Transform Methods

So far, we have just looked at EIS measurements in which the frequency is stepped over a range of several decades. An interesting alternative is to *apply several different frequencies at the same time* (Figure 5.14). This is the basis of the *multisine method* for impedance measurements, which is used for impedance measurements on batteries, bioimpedance and corrosion, to name a few examples. The objective is to achieve a shorter measurement time. Full details of the method can be found in the literature, but here we will just focus on the basic concept. The key issue is how to 'unscramble' the different frequencies in the signal and response. This might seem difficult, but in fact the digital signal processing in most commercial FRA systems allows you to do this using the *Fast Fourier Transform* (FFT) algorithm.

To illustrate how the Fourier transform unscrambles complex signals, I have taken just three sine waves with different frequencies and different phases and added them together as shown in Figure 5.15. The choice of random phases is important to avoid the sum of the signals giving large amplitudes. The choice of frequencies may seem odd, but on closer inspection, you should be able to see that they form a sequence $2\omega_0$, $3\omega_0$, $4\omega_0$. In other words, they are all multiples of a fundamental frequency (Figure 5.15).

You are probably familiar with the technique of Fourier transform infrared spectroscopy (FTIR). This involves 'unscrambling' an interferogram produced by a moving mirror in the spectrometer to obtain absorbance as a function of frequency (wavenumber). The FFT algorithm is an example of a discrete Fourier Transform (DFT). The FFT does the unscrambling job in multisine techniques. Applied to our simple example, it transforms the periodic signal in the *time domain* into three 'spikes' located at $2\omega_0$, $3\omega_0$, $4\omega_0$ in the *frequency domain*. These spikes are examples of the *Dirac delta function*. In fact, the three spikes shown represent the magnitude of the three sine waves that are present in the input signal. The

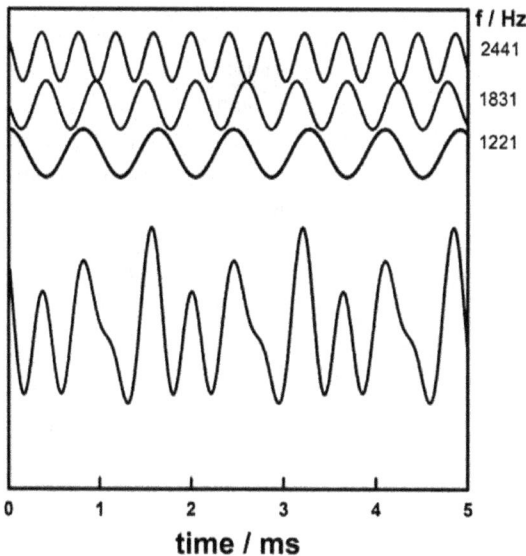

Figure 5.15. A simple example of a *multisine signal* in the time domain that we can use to test the Fourier transform. It contains three sine waves with different frequencies and phases added together.

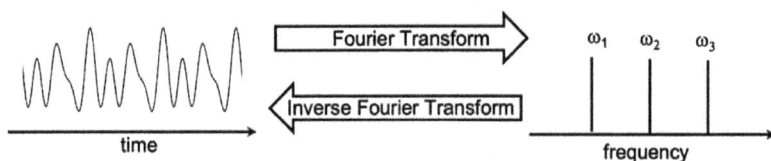

Figure 5.16. The Fourier transform and its inverse link the time and frequency domains. Here, the three-component multisine signal is transformed into three 'spikes' or Dirac delta functions in the frequency domain.

Fourier transform also provides information about the phase of each signal, either as a phase angle or as the following real and imaginary components. The inverse Fourier transform takes functions or data from the frequency domain and transforms them into the time domain (Figure 5.16).

We shall now take a look at a quick summary of the theory behind the method. First, we look at the way that Fourier series can be used to represent any periodic function.

Theory Note 5.1. Fourier Series

Jean Baptiste Fourier (1769–1830) was a busy man. He accompanied Napoleon on his ill-fated Egypt expedition and was probably the first to propose the greenhouse effect. Today, he is best known as the originator of the Fourier series and the Fourier transform.

The *infinite Fourier series* can be used to represent any periodic waveform in terms of sine and cosine terms with coefficients $a_1, a_2, \ldots, a_n$ for the cosine terms and $b_1, b_2, \ldots, b_n$ for the sine terms. If the *period* of the function is $2\pi/\omega_0$, the series looks as follows:

$$f(t) = \frac{1}{2}a_0 + a_1 \cos(\omega_0 t) + a_2 \cos(2\omega_0 t) + a_3 \cos(3\omega_0 t)$$

$$+ \cdots + a_n \cos(n\omega_0 t) + b_1 \sin(\omega_0 t) + b_2 \sin(2\omega_0 t)$$

$$+ b_3 \sin(3\omega_0 t) + \cdots + b_n \sin(n\omega_0 t) \qquad (5.1)$$

which can be written as the first term plus a sum of two separate series:

$$f(t) = \frac{1}{2}a_0 + \sum_{n=1}^{\infty} a_n \cos(n\omega_0 t) + \sum_{n=1}^{\infty} b_n \sin(n\omega_0 t) \qquad (5.2)$$

The term $\frac{1}{2}a_0$ corresponds to the *dc component* of the signal.

Theory Note 5.1. (*Continued*)

Here is a nice example of a Fourier series for a function of the variable x that includes only the cosine terms.

$$f(x) = \frac{2}{\pi} \left[\cos \omega_0 t - \frac{1}{3} \cos 3\omega_0 t + \frac{1}{5} \cos 5\omega_0 t \right.$$

$$\left. - \frac{1}{7} \cos 7x\omega_0 t + \frac{1}{9} \cos 9x\omega_0 - \cdots \right] \qquad (5.3)$$

Only the first nine terms have been added together in Figure 5.17, but you can see that the Fourier series represents a square wave with a peak-to-peak amplitude of 1 and zero dc component. Including more terms will remove the ripples. Note that the x axis corresponds to time normalized by dividing by the signal period, T, which is the inverse of the signal frequency, f. Zero time is at the centre of the square wave.

A more convenient way to represent the Fourier series is to convert the cosine and sine terms to *complex exponentials*. How do we do this? It's easy if we remember that $e^{j\theta} = \cos \theta + j \sin \theta$ and $e^{-j\theta} = \cos \theta - j \sin \theta$.

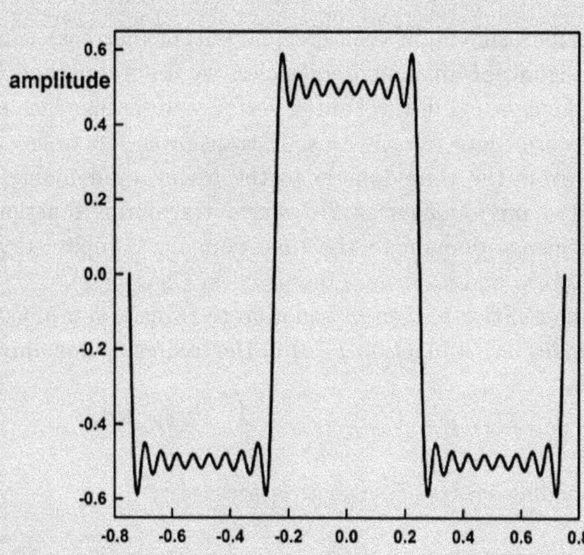

Figure 5.17. An example of a cosine Fourier series – equation (5.2) – that can be used to represent a periodic square wave signal. Only the first nine terms of the series have been computed.

(*Continued*)

Theory Note 5.1. (*Continued*)

A simple solution of these two simultaneous equations gives the following useful relationships:

$$\cos\theta = \frac{e^{j\theta} + e^{-j\theta}}{2} \qquad \sin\theta = \frac{e^{j\theta} - e^{-j\theta}}{2j} \tag{5.4}$$

If the sine and cosine terms in the Fourier series equation are replaced with their complex exponential equivalents, the general form of the Fourier series looks as follows:

$$f(t) = \sum_{n=-\infty}^{\infty} c_n e^{jn\omega_0 t} \tag{5.5}$$

where the Fourier coefficients c_n are now *complex quantities* given by

$$c_n = \frac{a_n - jb_n}{2} \tag{5.6}$$

This is a very neat form of the Fourier series because we no longer need to add the dc term a_0 – it is already defined by equation (5.5) for $n = 0$.

Since our multisine input (voltage) and output (current) signals consist of a sum of signals of different frequencies, we need a method to find the *complex coefficients*, c_n, in the Fourier series representing each signal. The Fourier transform does exactly this. It transforms functions – or in our case, *data* – from the time domain to the frequency domain (this works both ways: the *inverse Fourier Transform* transforms functions or data from the frequency domain to the time domain. Compare this with the *Laplace transform* and its inverse discussed in Chapter 1).

Here is the *Fourier transform* equation to transform a function $f(t)$ in the time domain into a function $F(\omega)$ in the frequency domain:

$$\mathcal{F}[f(t)] = F(\omega) = \frac{1}{\sqrt{2\pi}} \int_{-\infty}^{\infty} f(t)e^{-j\omega t} dt \tag{5.7}$$

The corresponding *inverse Fourier transform* is

$$\mathcal{F}^{-1}[F(\omega)] = f(t) = \frac{1}{\sqrt{2\pi}} \int_{-\infty}^{\infty} F(\omega)e^{j\omega t} d\omega \tag{5.8}$$

In order to see how the Fourier transform produces 'spikes' in the frequency domain, it is worth taking some time to look at the Dirac delta function. The next Theory Note follows this up.

Theory Note 5.2. The Dirac Delta Function and the Fourier Transform of $\cos(\omega_0 t)$

Paul Dirac was a brilliant English physicist who won the Nobel Prize in Physics in 1933. The delta function makes its first appearance in his 1930 book *The Principles of Quantum Mechanics*. He was famously taciturn; so much so that his Cambridge colleagues defined a Dirac unit as 1 word per hour.

The Dirac delta function is not strictly a function. It can be thought of as a limit of a distribution, for example, one looking like a 'top hat' (Figure 5.18).

Mathematically, this top hat shape is described as follows (TH stands for top hat):

$$\text{TH}(x) = \begin{cases} 0, & -\infty < x < -a/2 \\ \frac{1}{a}, & -\frac{a}{2} < x < +a/2 \\ 0, & +\frac{a}{2} < x < \infty \end{cases} \tag{5.9}$$

You can see that the area under the distribution is a multiplied by $1/a$, giving 1. Now let us see what happens as make a smaller – in fact tending towards zero. As the base narrows towards zero, the height will tend towards infinity, *but the area remains 1*. The Dirac delta function corresponds exactly to this limit. It is given the symbol $\delta(x)$, and it can be represented as a vertical arrow on an x axis labelled with 1. We can multiply the Dirac delta function by any number, real or complex, and the area will increase correspondingly. For real number multipliers, we represent this by increasing the height of the arrow. For complex number multipliers, we can either show real and imaginary arrows of appropriate lengths or magnitude and phase arrows. For our purposes, we will be

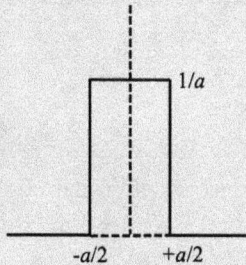

Figure 5.18. 'Top hat' function with constant area of 1. The Dirac delta function corresponds to the limit $a \to 0$.

(Continued)

Theory Note 5.2. (*Continued*)

using the Dirac delta function in the frequency domain to represent the spikes that appear at different frequencies in the Fourier-transformed signal.

To see how the delta function turns up when we carry out a Fourier transform, let us take a cosine wave $f(t) = \cos(\omega_0 t)t$ and input it into the Fourier transform equation as follows:

$$\mathcal{F}[f(t)] = F(\omega) = \frac{1}{\sqrt{2\pi}} \int_{-\infty}^{\infty} f(t)e^{-j\omega t}dt = \frac{1}{\sqrt{2\pi}} \int_{-\infty}^{\infty} \cos(\omega_0 t)e^{-j\omega t}dt \tag{5.10}$$

We can convert the cosine term into complex exponential form by remembering that $e^{j\theta} = \cos\theta + j\sin\theta$ and $e^{-j\theta} = \cos\theta - j\sin\theta$. If we add these two equations, we get $2\cos\theta = e^{j\theta} + e^{-j\theta}$. We can therefore use the complex exponential form of $\cos(\omega_0 t)$:

$$\cos(\omega_0 t) = \frac{e^{j\omega_0 t} + e^{-j\omega_0 t}}{2} \tag{5.11}$$

to obtain

$$\mathcal{F}[\cos(\omega_0 t)] = \frac{1}{2\sqrt{2\pi}} \left[\int_{-\infty}^{\infty} e^{j\omega_0 t}e^{-j\omega t}dt + \int_{-\infty}^{\infty} e^{-j\omega_0 t}e^{-j\omega t}dt \right] \tag{5.12}$$

This simplifies to

$$\mathcal{F}[\cos(\omega_0 t)] = \frac{1}{2\sqrt{2\pi}} \left[\int_{-\infty}^{\infty} e^{j(\omega-\omega_0)t}dt + \int_{-\infty}^{\infty} e^{-j(\omega+\omega_0)t}dt \right] \tag{5.13}$$

The integrals of these complex exponentials vanish except when $\omega - \omega_0 = 0$, when the two terms become Dirac delta functions that tell us that we will have spikes located *at* $\omega = \omega_0$ and at $\omega = -\omega_0$.

$$\mathcal{F}[\cos(\omega_0 t)] = \frac{1}{2\sqrt{2\pi}} [2\pi\delta(\omega - \omega_0) + 2\pi\delta(\omega + \omega_0)] \tag{5.14}$$

which simplifies to

$$\mathcal{F}[\cos(\omega_0 t)] = \sqrt{\frac{\pi}{2}} [\delta(\omega - \omega_0) + \delta(\omega + \omega_0)] \tag{5.15}$$

Note that the Fourier transform of cosine has no terms involving j, so that it is real. This means that if we use polar coordinates, i.e., magnitude and phase, the *phase angle is zero*.

Theory Note 5.2. (*Continued*)

Fourier Transform of cosine   Fourier Transform of sine

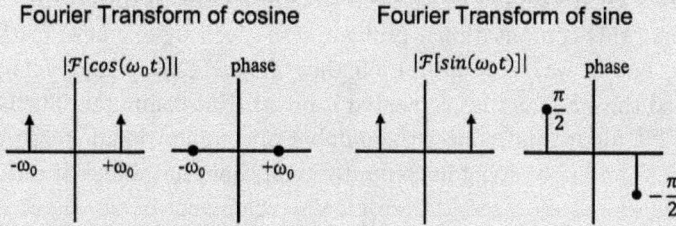

Figure 5.19. Fourier transformation of cosine and sine functions gives Dirac delta functions. The figure shows the magnitude and phase of $\mathcal{F}[\cos(\omega_0 t)]$ and $\mathcal{F}[\sin(\omega_0 t)]$ located at $-\omega_0$ and $+\omega_0$.

If the same approach is taken to find the Fourier transform of $\sin(\omega_0 t)$, we start with

$$\sin(\omega_0 t) = \frac{e^{j\omega_0 t} - e^{-j\omega_0 t}}{2j} \tag{5.16}$$

and find that there are again two solutions. However, both are now *imaginary*, and one is positive and the other negative.

$$\mathcal{F}[\sin(\omega_0 t)] = j\sqrt{\frac{\pi}{2}}[\delta(\omega - \omega_0) - \delta(\omega + \omega_0)] \tag{5.17}$$

This means that we have a phase angle of $+\pi/2$ for the first term and $-\pi/2$ for the second. We can represent these two results using polar coordinates (magnitude and phase) (Figure 5.19).

In practice, we have data rather than a function to input into the Fourier transform. We collect data in the form of individual data points over a finite time span (*data sampling*). In the FRA, analogue to digital converters sample the signals at discrete time intervals Δt to give samples $1, \ldots, N$ at times 0, Δt, $2\Delta t$, $3\Delta t, \ldots, n\Delta t$. The total time is therefore $(N-1)\Delta t$, where N is the total number of samples. To analyze this data, the integral in the Fourier transform is replaced by a summation in the *Discrete Fourier Transform* (DFT), which looks as follows:

$$x_k = \sum_{n=0}^{N-1} x_n e^{-j\frac{2\pi k n}{N}} \tag{5.18}$$

where $k = 0, \ldots, N-1$.

You can see that evaluation of the DFT appears to require N^2 steps because for each value of k, we need to sum for $n = 0$ to $n = N - 1$. The solution to this computational problem was given by our hero Carl Gauss in 1805, but it was not until 1965 that *James Cooley* and *John Tukey* developed the FFT and implemented it on an IBM computer. The Cooley–Tukey FFT algorithm reduces the number of computational steps from N^2 to $N \log N$ – a huge saving in computational time that makes it possible to handle enormous data sets (the original development of the algorithm was in response to the need to analyze the output of multiple sensors set up to detect seismic waves from nuclear tests in the Soviet Union).

The application range of the FFT is enormous, ranging from engineering and science to music, and implementations are readily available in all modern computer languages and in everyday spreadsheets like Microsoft Excel. The FFT can be used in one, two or three dimensions, and FFT image processing is used on length scales ranging from picometers (scanning tunnelling microscopy) to light years (James Webb Space Telescope). In the

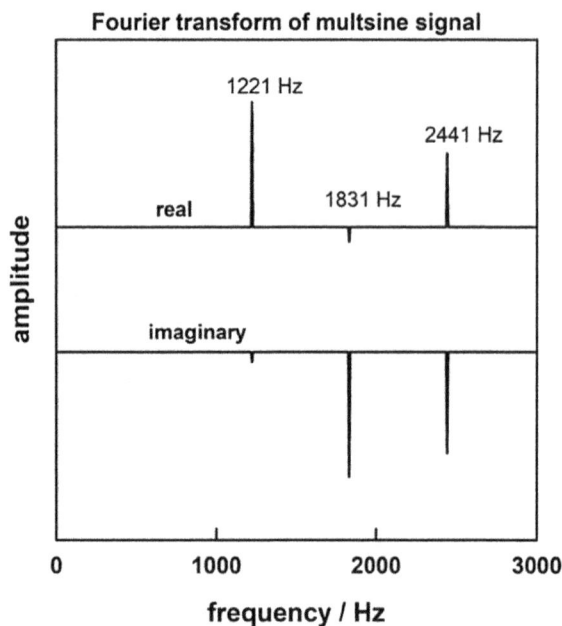

Figure 5.20. Output of FFT of the three-sine composite signal in Figure 5.15. Note that the output shows the real and imaginary components of the three signals, allowing the magnitudes and relative phases to be calculated.

much more down-to-earth example in Figure 5.15, three different sine waves with different phases and different frequencies were added to produce a *multisine signal.* This data has been used to test the FFT procedure using a spreadsheet programme. The transform takes a block of 2^n data points, and the output is in the form (real, imaginary) rather than in polar coordinates, but of course, it is easy to interconvert the two. Figure 5.20 shows the result. In the multisine method, the FFT is used to obtain the voltage and current at the chosen frequencies, allowing calculation of the impedance.

References

Analog Devices. *Operational Amplifiers (Op Amps).* Analog Devices. Available: https://www.analog.com/en/product-category/operational-amplifiers.html Accessed 4th July 2023.

Cogger, N. D. & Evans, N. J. 1999. *Technical Report No 6. An Introduction to Electrochemical Impedance Measurement.* Solartron. Available: https://www. ameteksi.com/library/application-notes/solartron-analytical Accessed 4th July 2023.

Tran, A. T., Huet, F., Ngo, K. & Rousseau, P. 2011. Artefacts in electrochemical impedance measurement in electrolytic solutions due to the reference electrode. *Electrochimica Acta,* 56, 8034–8039.

Chapter 6

Examples of Finite Diffusion Impedance: The Rotating Disc Electrode and the Ultramicroelectrode

6.1 Introduction

In the first part of this chapter, we look at some real-life experimental data obtained in our Bath laboratory by my colleague Professor Frank Marken using a rotating disc electrode. The *finite Warburg diffusion impedance* of this electrode was discussed in Chapter 3. I recommend that before continuing you revisit Chapter 3, Section 3.6, which discusses the finite Warburg impedance for the rotating disc electrode. We have chosen the hexacyanoferrate redox couple $Fe(CN)_6^{4-}/Fe(CN)_6^{3-}$ because it is often considered a model system with relatively fast electron transfer kinetics. For the rotating disc electrode, the focus is on the effect of rotation rate on the width of the finite diffusion layer and how this determines the shape of the low-frequency part of the impedance response. We show how analysis of the impedance as a function of frequency gives information about the averaged diffusion coefficient of the $Fe(CN)_6^{3-/4-}$ redox ions. We also look at how applying an overpotential affects the impedance signature.

We also look at the same system using an *ultramicroelectrode* (UME). We examine how the microdisc geometry affects the diffusion of species to and from the electrode surface and explain why we cannot use the standard finite Warburg impedance to model the impedance response.

6.2 Experimental Details

The electrolyte used in these experiments consisted of a 5 mM $K_4Fe(CN)_6$ + 5 mM $K_3Fe(CN)_6$ in 1.0 M KCl. The rotating disc setup and details of the electrode construction are illustrated in Figure 6.1.

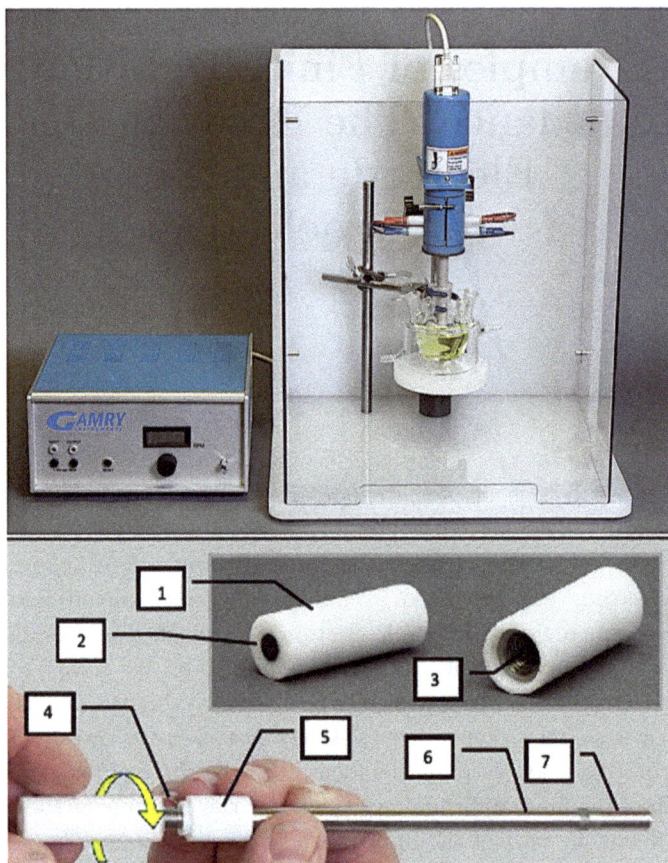

Figure 6.1. Rotating disc setup showing cell, rotator and RDE control used for the experiments with the hexacyanoferrate redox system. The lower part of the figure shows details of the electrode. (1) PTFE sheath. (2) Platinum electrode. (3, 4) Screw connection to rotor. (5) Upper PTFE sheath. (6) Drive shaft. (7) Connection point for potentiostat via silver-loaded carbon brushes. Actually, the electrode pictured is a *ring-disc* electrode. You may be able to see the ring if you look closely. The ring disc electrode is used to study the products of reactions occurring on the disc. Only the disc was connected in these experiments. Adapted with kind permission from Gamry Instruments.

6.3 A Closer Look at the Rotating Disc Electrode

We start by looking at the theoretical current voltage plot calculated for a rotating disc electrode in the case where we have equal concentrations of oxidized and reduced redox species. For a fixed rotation rate at the RDE, the current voltage plot for a 'reversible' redox system (i.e., one in which the electrode process is diffusion controlled) can be expressed in terms of the anodic and cathodic limiting current densities, $j_{l,a}$ and $j_{l,c}$, by the *voltammetric wave equation*.

$$E = E_{1/2} + \frac{RT}{nF} \ln \frac{j_{l,c} - j}{j - j_{l,a}} \tag{6.1}$$

Here, $E_{1/2}$ is called the *halfwave potential*, which is the potential at which the current is halfway between the cathodic and anodic limits, i.e., when $j = \frac{j_{l,a} + j_{l,c}}{2}$.

Equation (6.1) is not particularly handy for calculating the current voltage plot, so let us rearrange it and take the exponential of each side of the equation to get rid of the natural logarithm term. We obtain

$$j = \frac{j_{l,c} + j_{l,a} \left\{ \exp \left[\frac{nF}{RT} \left(E - E_{1/2} \right) \right] \right\}}{\left\{ 1 + \exp \left[\left[\frac{nF}{RT} \left(E - E_{1/2} \right) \right] \right] \right\}} \tag{6.2}$$

The anodic and cathodic limiting current densities $j_{l,a}$ and $j_{l,c}$ depend on the rotation rate ω_{rot} of the RDE and are given by the *Levich equation*.

$$\lfloor j_l \rfloor = 0.62 n F D^{2/3} \nu^{-1/6} \omega_{\text{rot}}^{1/2} C \tag{6.3}$$

The rotation rate ω_{rot} in equation (6.3) is expressed in *radians* s^{-1}, i.e., as 2π times the rotation frequency in Hz. Here, $D = D_O$ for the cathodic limiting current density (negative) and $D = D_R$ for the anodic limiting current density (positive). $C = C_O$ or C_R as appropriate (note that the concentration units in the Levich equation are mol cm^{-3}). The ν term in equation (6.4) is the *kinematic viscosity* (viscosity/density) of the electrolyte. For 1 M KCl at 20°C, its value is close to 10^{-2} cm^2 s^{-1}. Figure 6.2 shows the shape of the current potential plot predicted by equation (6.2).

At this point you may be wondering how the Levich equation relates to the diffusion profile. The diffusion profiles for the anodic limiting current are illustrated in Figure 6.3. Figure 6.4 gives you an idea of how wide the diffusion layer is at different rotation rates. The rotation rates used in this experimental study are marked on the plot.

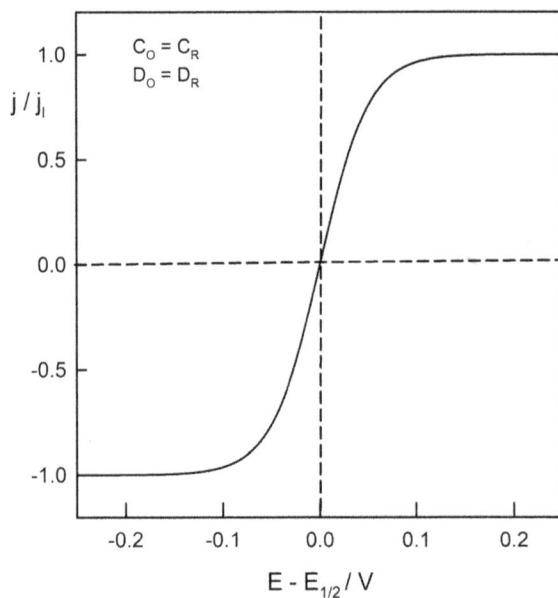

Figure 6.2. Normalized theoretical current voltage plot calculated from equation (6.3) for 'reversible', i.e., diffusion-controlled, redox system. $n = 1$, $C_O = C_R$ and $D_O = D_R$. In this particular case, $E_{1/2}$ is equal to the *formal potential*. Normalization involves dividing the currents by the limiting current.

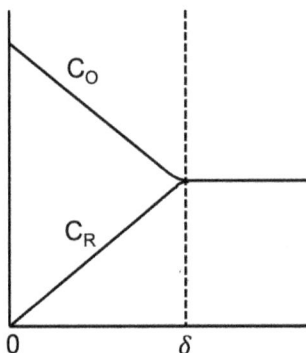

Figure 6.3. Diffusion profiles under limiting anodic current conditions. The current is determined by the concentration gradient C_R/δ_{RDE}, where δ_{RDE} depends on the inverse square root of the rotation rate ω_{rot}.

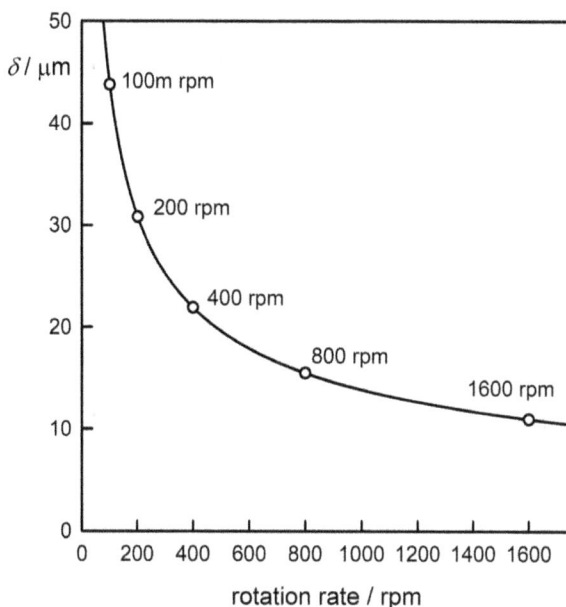

Figure 6.4. Plot showing how the diffusion layer thickness shrinks as the rotation rate is increased, reflecting the dependence of δ_{RDE} on $\omega_{\text{rot}}^{1/2}$. The marked points indicate the rotation rates used in the experiments.

Remember that the concentration *ratio* of oxidized and reduced species *at the electrode surface* is given by the Nernst equation if electron transfer is fast, so the surface concentration of the reduced species is very low in the limiting anodic current region. It follows that the limiting anodic current density $j_{L,a}$ is given by Fick's First Law.

$$j_{L,a} = nF \left. \frac{dC_R}{dx} \right|_{x=0} = nF \frac{C_R^{\text{bulk}}}{\delta} \tag{6.4}$$

If we compare this with the Levich equation, we see that δ for the RDE must be given by

$$\delta_{\text{RDE}} = 1.61 D^{-2/3} \nu^{1/6} \omega_{\text{rot}}^{-1/2} \tag{6.5}$$

We are going to need δ_{RDE} when we come to discuss the finite Warburg diffusion impedance.

If we use the Levich equation for the limiting currents, we find that the halfwave potential $E_{1/2}$ is given by

$$E_{1/2} = E^{0'} + \frac{RT}{nF} \ln \left(\frac{D_R}{D_O} \right)^{\frac{2}{3}} \tag{6.6}$$

Here, D_R and D_O are the diffusion coefficients of the oxidized and reduced species, respectively, and $E^{0'}$ is the *formal potential*, i.e., the equilibrium potential measured vs. the standard hydrogen electrode when $C_O = C_R$. If $D_O = D_R$, we see that $E_{1/2} = E^{0'}$, as is the case in Figure 6.2.

6.4 Analyzing the Experimental Results for the RDE

We can now compare the predicted current voltage plots with the experimental RDE results shown in Figure 6.5.

We can see that the limiting currents scale nicely with $\omega_{rot}^{1/2}$, although the cathodic limiting current is not perfectly flat, but shows some additional current as we scan to more negative potentials. This is probably due to the reduction of dissolved oxygen, because the solution was not degassed.

Figure 6.6 shows the Levich plots – see equation (6.3) – of the cathodic and anodic limiting currents measured at 0 V and 0.5 V vs. SCE, respectively. The diffusion coefficients of the oxidized and reduced species determined from the slope of these plots are 6.8×10^{-6} cm^2 s^{-1} for $Fe(CN)_6^{4-}$ and 7.3×10^{-6} cm^2 s^{-1} for $Fe(CN)_6^{3-}$. The diffusion coefficients

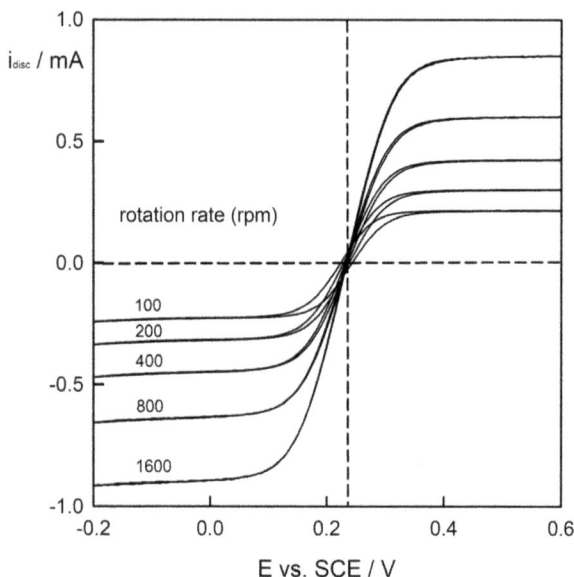

Figure 6.5.   Current voltage plots for the RDE obtained at different rotation rates (rpm = revolutions per minute of the RDE) for 5 mM $K_4Fe(CN)_6$+ 5 mM $K_3Fe(CN)_6$ in 1.0 M KCl. Potentials vs. SCE (saturated calomel electrode). Radius of platinum RDE 3 mm.

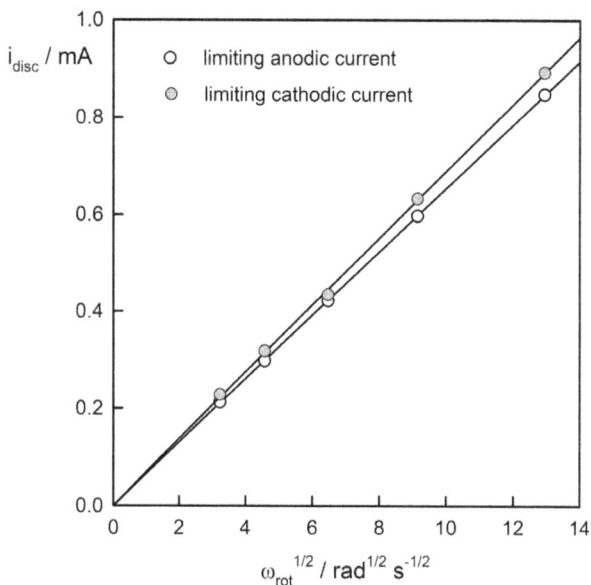

Figure 6.6. Levich plots of the anodic and cathodic limiting currents shown in Figure 6.4. The slopes are used to determine the diffusion coefficients for the oxidized and reduced species as described in the text.

of the two species are not identical because on average the more highly charged $Fe(CN)_6^{4-}$ is associated with more K^+ ions than $Fe(CN)_6^{3-}$, increasing its hydrodynamic radius.

The first set of impedance measurements was recorded at the equilibrium potential using the same rotation rate sequence as shown in the current voltage plots in Figure 6.5. Figure 6.7 shows the Nyquist plots for three rotation rates, 100, 400 and 1600 rpm. Note how the low-frequency parts of the response correspond to the shape expected for a finite Warburg impedance (see Chapter 3, Section 3.6).

At this point let us remind ourselves of the expression for the finite Warburg impedance associated with diffusion across the diffusion layer set up by the RDE. This is given by equation (3.15) in Chapter 3. Here it is again:

$$Z_{W,s} = \frac{Z_{W,s}(0)\tanh(j\Lambda)^{0.5}}{(j\Lambda)^{0.5}} \tag{6.7}$$

where $\Lambda = \omega\delta^2/D$. Be careful not to confuse ω, the ac radial frequency, with ω_{rot}, the rotation rate of the RDE.

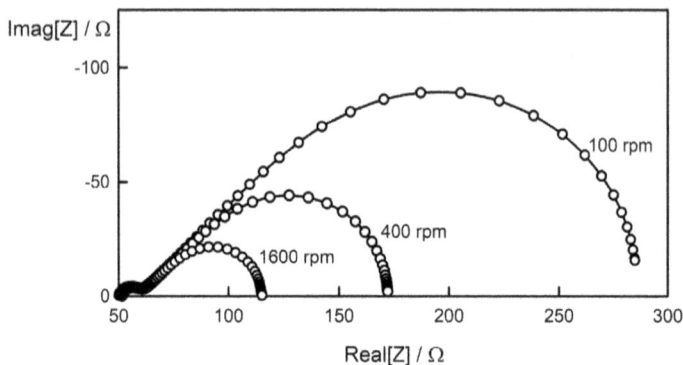

Figure 6.7.   Nyquist plots showing how the impedance of the hexacyanoferrate system varies with rotation rate. Note that the low-frequency intercept of the finite Warburg impedance should scale with $1/\omega_{\mathrm{rot}}^{1/2}$.

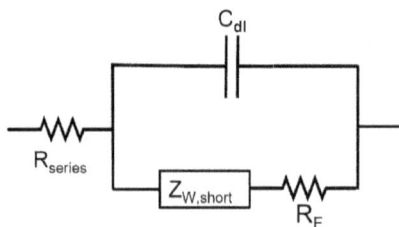

Figure 6.8.   Modified Randles equivalent circuit used for analyzing the impedance measurements made with the RDE. The semi-infinite Warburg impedance in the original Randles circuit has been replaced by a finite Warburg impedance.

The low-frequency limit of the Warburg impedance is real and is given by

$$Z_{W,s}(0) = \frac{2RT\delta_{\mathrm{RDE}}}{An^2F^2CD} \tag{6.8}$$

The key variable in these expressions is δ_{RDE}, the thickness of the diffusion layer, which we know from the Levich equation, depends inversely on the square root of the rotation rate. It follows that the low-frequency intercept on the real axis of the Nyquist plots in Figure 6.7 should scale with $\omega_{\mathrm{rot}}^{-1/2}$.

The impedance data obtained as a function of rotation rate were fitted in ZView® using the modified Randles equivalent circuit shown in Figure 6.8.

An example of the fitting is shown in Figure 6.9 for the impedance data measured at the equilibrium potential at a rotation rate of 100 rpm. To obtain a better fit, the double-layer capacitance was replaced by

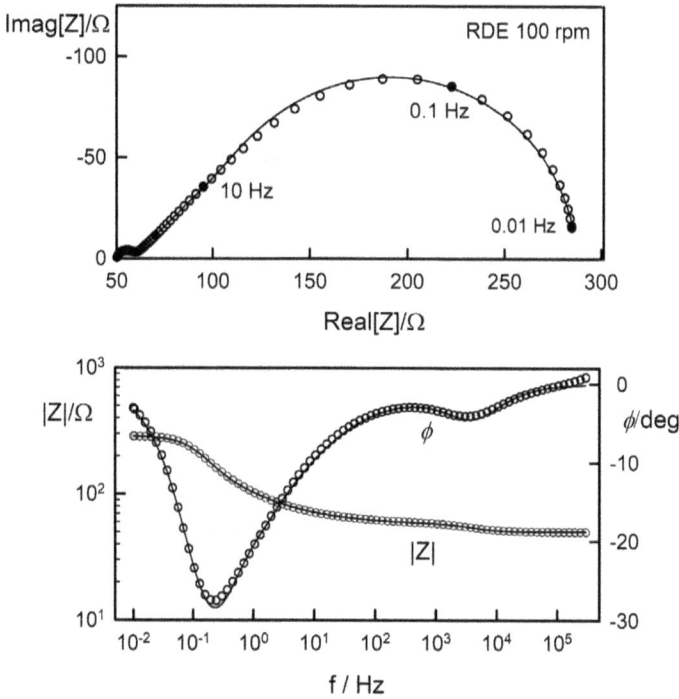

Figure 6.9. Nyquist and Bode plots illustrating the ZView® fitting used to obtain parameter values of the finite Warburg impedance. The data analyzed in this example were obtained at a rotation rate of 100 rpm. Points – experimental. Line – fit.

a constant phase-shift element to account for non-ideal behaviour of the platinum electrode. The same fitting procedure was used for the data obtained at the other rotation rates, and Figure 6.10 confirms the expectation that $Z_{W,s}(0)$ should scale with $\omega_{\text{rot}}^{-1/2}$.

Looking back at Chapter 3, Section 3.6, you will see that ZView® also fits the *diffusion time constant* τ_{diff} (labelled T in ZView®). This time constant is given by

$$\tau_{\text{diff}} = \frac{\delta^2}{D} \tag{6.9}$$

where D is a mean diffusion coefficient and δ is the width of the diffusion layer (δ_{RDE}), which, in the case of the RDE, depends on $\omega_{\text{rot}}^{-1/2}$. Since the δ term is squared in equation (6.9), we expect the diffusion time constant to depend on $1/\omega_{\text{rot}}$. A log–log plot of $\tau_{\text{diff}}(T$ in ZView®) vs. ω_{rot} should

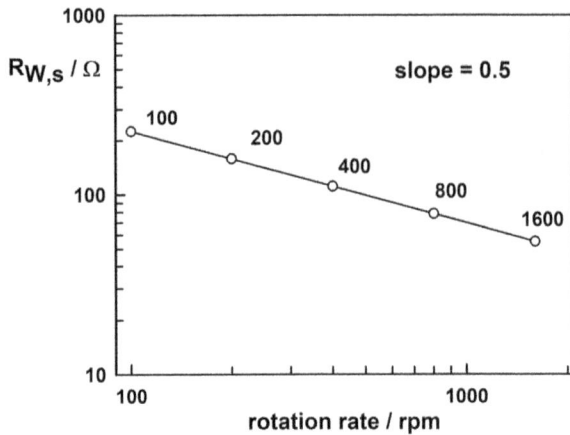

Figure 6.10. Log–log plot of $Z_{W,s}(0)$ vs. ω_{rot}. The slope of $-1/2$ confirms that $Z_{W,s}(0)$ depends inversely on the square root of the rotation rate. This variation reflects the dependence of the diffusion layer thickness on rotation rate – see equations (6.5) and (6.8).

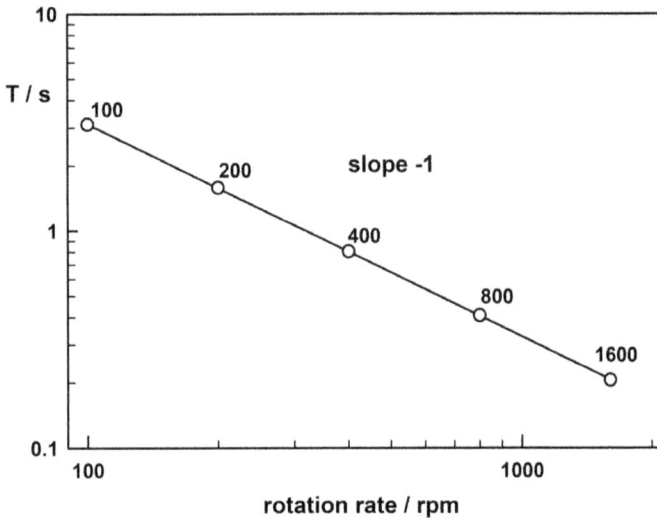

Figure 6.11. Log–log plot of diffusion time constant vs. rotation rate. The slope of -1 confirms that τ_{diff} depends inversely on the rotation rate ω_{rot}.

therefore have a slope of -1. This is indeed the case for our experimentally obtained diffusion time constants as shown by Figure 6.11.

Remember that the diffusion time constant is given by equation (6.9). Since equation (6.5) shows that δ^2 depends linearly on ω^{-1}, we can now

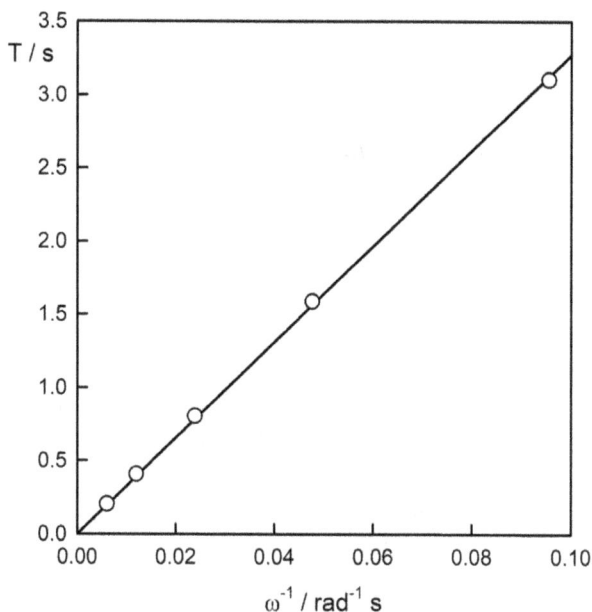

Figure 6.12. Plot of diffusion time constant T vs. ω^{-1}. The *mean diffusion coefficient* is determined from the slope of the plot.

obtain the mean diffusion coefficient from a plot of T vs. ω^{-1}. From the slope of the plot in Figure 6.12 we find that the mean diffusion coefficient of the hexacyanoferrate system is 6.2×10^{-6} cm^2 s^{-1}, which is reasonably close to the values obtained from the Levich plots.

Now let us see if we can relate the 'dc' value of the diffusion impedance to the current voltage plot in some way. If the redox system is 'reversible', i.e., electron transfer is fast, the ratio C_O/C_R at the surface of the electrode is determined by the Nernst equation for every point on the current voltage plots. What does this mean for the diffusion impedance? In the zero-frequency limit of the diffusion impedance, the diffusion resistance is determined by the inverse of the slope of the j–E plot (this is simply Ohm's law $V = IR$ applied to a small section of the j–E plot).

$$Z_{W,s}(0) = R_W = \frac{dE}{dj} \tag{6.10}$$

To see how we can obtain dE/dj, we need a Theory Note.

Theory Note 6.1. Finding the Diffusion Impedance

Look back at equation (6.2). We would like to use it to find the derivative dE/dj. We will do this by finding dj/dE and inverting the result. For convenience, here is equation (6.2) again.

$$j = \frac{j_{l,c} + j_{l,a}\left\{\exp\left[\frac{nF}{RT}\left(E - E_{1/2}\right)\right]\right\}}{\left\{1 + \exp\left[\frac{nF}{RT}\left(E - E_{1/2}\right)\right]\right\}}$$

Let us simplify this a bit by using the Greek letter χ (chi) to represent the dimensionless exponential term $\exp\left[\frac{nF}{RT}(E - E_{1/2})\right]$, so that equation (6.2) becomes

$$j = \frac{j_{l,c} + \chi j_{l,a}}{(1 + \chi)} \quad \text{where } \chi = \exp\left[\frac{nF}{RT}(E - E_{1/2})\right] \tag{6.11}$$

You might remember the *chain rule* for differentiation of a quotient:

$$d\left(\frac{u}{v}\right) = \frac{v\,du - u\,dv}{v^2} \tag{6.12}$$

Let us represent the derivative of χ with respect to E as χ'. Then our solution can be written as

$$\frac{dj}{dE} = \frac{(1 + \chi)j_{l,a}\chi' - (j_{l,c} + \chi j_{l,a})\chi'}{(1 + \chi)^2} \tag{6.13}$$

The derivative χ' with respect to E is given by

$$\chi' = \frac{nF}{RT}\exp\left(\frac{nF}{RT}(E - E_{1/2})\right) \tag{6.14}$$

Substituting this into equation (6.8) and inverting the results gives us the *diffusion resistance* as a function of potential $(E - E_{1/2})$.

$$\frac{dE}{dj} = Z_{W,s}(0) = R_W = \frac{\left\{1 + \exp\left[\frac{nF}{RT}(E - E_{1/2})\right]\right\}^2}{\left\{\frac{nF}{RT}\exp\left[\frac{nF}{RT}(E - E_{1/2})\right]\right\}(j_{l,a} - j_{l,c})} \tag{6.15}$$

We will come back to equation (6.15) after looking at some more experimental data.

The impedance measurements discussed so far were all made at the equilibrium potential, where no dc current flows. However, after the Theory Note, we are also able to analyze impedance data obtained at potentials away from E_{eq}. Impedance measurements were made at a rotation rate of 200 rpm and positive overpotentials of 59 mV and 118 mV. Why these potentials? The clue is the Nernst equation, which can be written in terms of $E^{0'}$, the formal potential using logarithms to base 10 as

$$E_{eq} = E^{0'} + \frac{2.303RT}{nF} \log_{10} \frac{C_O}{C_R} \tag{6.16}$$

When discussing the finite Warburg impedance, we have assumed that the concentration ratio C_O/C_R *at the surface of the electrode* ($x = 0$) is determined by the Nernst equation. This means that equation (6.16) still applies at any potential on the current potential plot. Rearranging the equation and substituting $n = 1$ for the hexacyanoferrate redox couple gives

$$\log_{10}\left(\frac{C_O}{C_R}\right) = \frac{RT}{2.303F}(E - E^{0'}) = \frac{E - E^{0'}}{59\text{mV}} \tag{6.17}$$

So, for $E - E^{0'} = 59\,\text{mV}$, $\log_{10}(C_O/C_R) = 1$, i.e., $C_O/C_R = 10$. For $E - E^{0'} = 118\,\text{mV}$, $\log_{10}(C_O/C_R) = 2$, i.e., $\frac{C_O}{C_R} = 100$.

After this short digression we will look at the results of fitting the impedance obtained at 200 rpm and three values of $E - E^{0'}$: 0 mV, 59 mV and 118 mV. Figure 6.13 shows that increasing the overpotential has a large effect on the finite Warburg impedance arc in the Nyquist plots. This reflects the fact that the slope of the $j - E$ plot decreases rapidly as the potential is moved away from equilibrium (see, for example, Figure 6.2).

The fitting gave three values of the zero-frequency finite Warburg impedance, which have been plotted in Figure 6.14 along with the predicted potential dependence obtained using equation (6.15). Starting

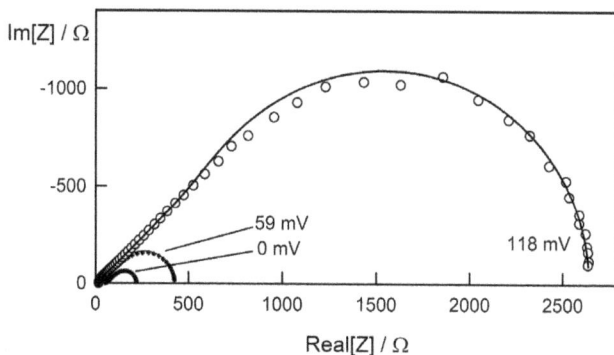

Figure 6.13. Nyquist plots of hexacyanoferrate system at the RDE recorded as a function of overpotential. Rotation rate 200 rpm. Note how the low-frequency arc expands as the overpotential is increased.

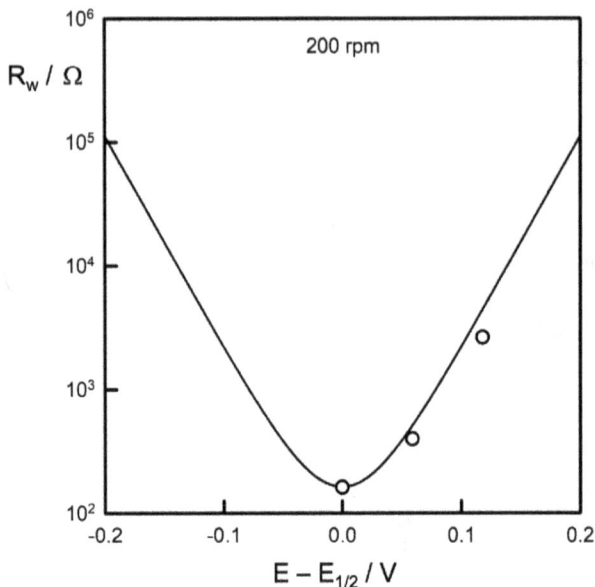

Figure 6.14. Potential dependence of the zero-frequency Warburg impedance R_W predicted by equation (6.15) for 200 rpm (line) compared with the values obtained by analyzing the impedance responses.

with equation (6.15), you should be able to prove that the magnitude of the slopes approach 59 mV/decade on each side of the plot.

6.5 When Is an Electrode an Ultramicroelectrode?

The obvious answer to this question is – when the electrode is really small. However, this leads to the second question – small relative to what? We need some sort of scale. The relevant scale is related to the width of the diffusion layer. We have seen that the RDE controls δ, the width of the diffusion layer via the rotation rate. If you look back at Figure 6.4, you will see that δ_{RDE} was in the range 10–50 microns in our experiments. But what would δ be if we didn't rotate the electrode? This is not an easy question to answer since we cannot establish a steady state if the RDE is not rotating. If the RDE is stationary and we run a cyclic voltammogram starting at a potential positive from the equilibrium potential, we see the response shown in Figure 6.15.

The steep decay when the potentiostat is switched on shows that the diffusion layer thickness increases with time so that the concentration

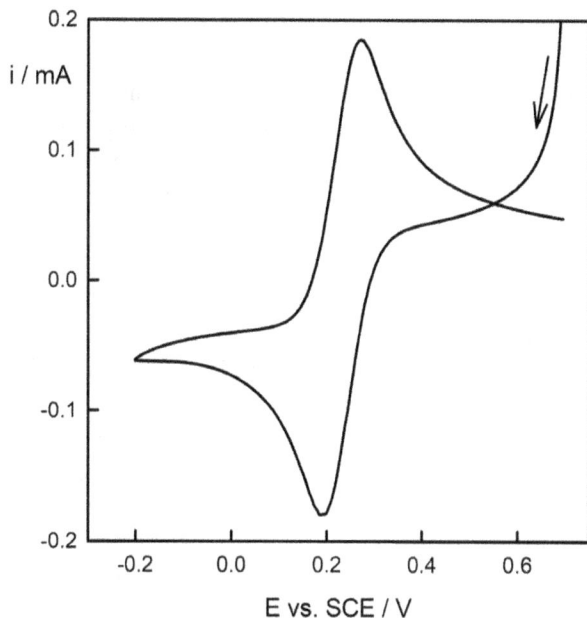

Figure 6.15. Current voltage plot obtained for the stationary RDE in 5 mM $K_4Fe(CN)_6$ + 5 mM $K_3Fe(CN)_6$ in 1.0 M KCl. The potential was scanned from the anodic limit to the cathodic limit and back again at a sweep rate of 20 mV s^{-1}. Note the initially high current when the potentiostat is switched on. The voltammogram then shows a cathodic peak followed by an anodic peak. Clearly, there is no steady-state diffusion under these circumstances. Instead. the diffusion layer thickness changes with time during the scan.

gradients of $Fe(CN)_6^{4-}$ and $Fe(CN)_6^{3-}$ at the surface become less steep. If we had simply switched on the potential and not scanned, the anodic current $i(t)$ would have fallen according to the *Cottrell equation*, which we saw in Chapter 3.

$$i(t) = \frac{nFAD^{1/2}C}{\pi^{1/2}t^{1/2}}$$ (6.18)

The Cottrell equation shows that the diffusion layer thickness increases with the square root of time. Fick's first law tells us that

$$i(t) = nFAD \left. \frac{dC}{dx} \right|_{x=0} = nFAD \frac{C^\infty}{\delta(t)}$$ (6.19)

It is easy to show from equations (6.18) and (6.19) that

$$\delta(t) = \sqrt{\pi Dt}$$ (6.20)

So, how large is the diffusion layer after, say, 100 ms? Using an order of magnitude value $D = 10^{-5}$ cm^2 s^{-1} gives $\delta(100\,\text{ms}) = 17.7$ microns. After 10 s, $\delta(10\,\text{s}) = 177$ microns. Returning to our original questions, we now have a scale on which to decide whether an electrode is small or large compared with $\delta(t)$. We are going to look at some results obtained using a 25-micron diameter disc electrode. This is clearly small on the time scale of seconds but not milliseconds.

The ultramicroelectrode that was used in the experiments with the hexacyanoferrate system was made by carefully sealing a 25-micron diameter platinum wire into a narrow bore soda glass tube under vacuum using a glassblowing torch. The exposed end of the wire was cut off and the resulting platinum microdisc was polished on successively smaller grades of alumina to give a mirror finish. Figure 6.16 shows scanning electron micrographs of the end surface of this type of electrode at two different magnifications. The higher magnifications show the electrode geometry, which is referred to as an *inlaid disc*.

Diffusion to the very small disc electrode takes place *radially* as illustrated in Figure 6.17. This type of diffusion is termed *hemispherical* and contrasts with the usual *linear* diffusion to a large planar electrode. The interesting thing about this diffusion geometry is that it sets up a *steady state* after some time. This may seem odd at first because we just saw in Chapter 3 that the current at a flat electrode falls with $t^{-1/2}$. However, in that case diffusion was *linear* because δ is much smaller than the electrode dimensions. The diffusion zone around the ultramicroelectrode can be thought of as analogous to the layers on an onion. As we go further away

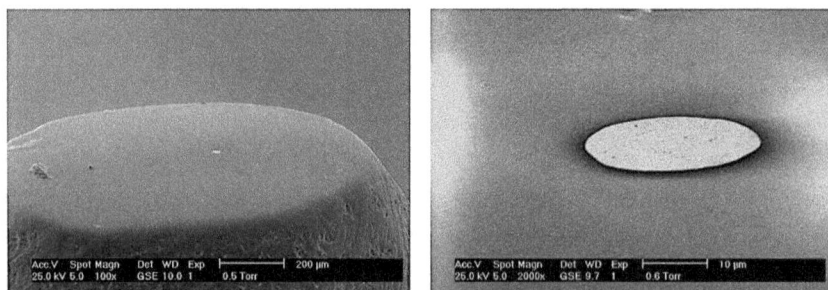

Figure 6.16. Scanning electrode microscope (SEM) images of platinum disc microelectrode of the type used in this study. Images kindly provided by Guy Denault (University of Southampton: the use of ultramicroelectrodes was pioneered at the University of Southampton by the late Professor Martin Fleischmann).

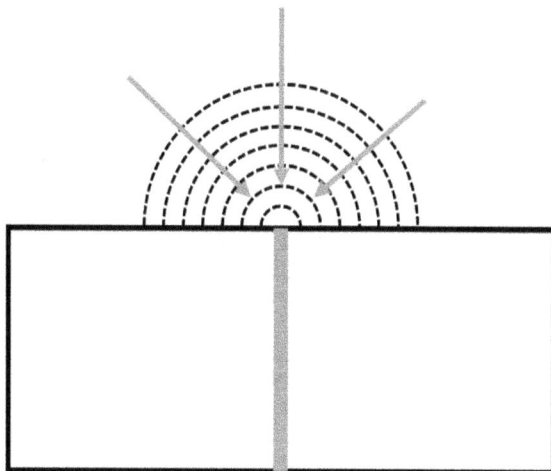

Figure 6.17. Steady-state hemispherical diffusion to an inlaid disc ultramicroelectrode. The dashed lines represent hemispherical shells of equal concentration.

from the centre, the shells have a larger surface area. It turns out that under steady-state conditions, the total amount diffusing into any shell matches the amount diffusing out. Suppose we have a hemispherical shell of thickness δr located at some radial distance r from the centre of the UME. Its volume is equal to the area of a hemisphere $(4\pi r^2)$ multiplied by the thickness δr of the shell, giving $4\pi r^2 \delta r$. The next shell further out is located at $r + \delta r$, so its volume is larger because it is $4\pi(r + \delta r)^2 \delta r \simeq 4\pi(r^2 + 2r\delta r)\delta r$. On the other hand, the next shell further in is located at $r - \delta r$, so its volume is smaller because it is $4\pi(r - \delta r)^2 \delta r \simeq 4\pi(r^2 - 2r\delta r)\delta r$. Solution of Fick's laws shows that the total flux into any hemispherical shell is exactly balanced by the flux out as a consequence of the dependence of the shell volume on distance from the centre of the UME.

This brings us to a second surprise. Normally, we would expect the current at an electrode to scale linearly with the electrode area. In other words, the current *density* j is expected to be independent of electrode size. This is true for semi-infinite linear diffusion, but not for hemispherical diffusion. Solution of Fick's laws for this geometry shows that limiting steady-state current density is given by

$$j = \frac{4nFDC}{\pi r_0} \tag{6.21a}$$

Since the area of the circular microdisc is $4\pi r^2$, we see that the current scales with r_0 rather than with r_0^2, as we might expect.

$$i = 4nFDCr_0 \qquad (6.21b)$$

If you look at equation (6.21a), you can see that the UME radius r_0 plays the same role as the thickness δ of the RDE diffusion layer. So, by making r_0 small, we can increase the current density. In other words, making the UME smaller has the same effect as rotating the RDE faster. However, there is a limit to how fast we can rotate the RDE before turbulence sets in and Levich's nice hydrodynamic solution no longer applies. By contrast, we can make UMEs really small. Silver-coated 5-micron platinum wire is available and has been used to fabricate glass-sealed UMEs. This wire is called Wollaston wire after William Hyde Wollaston, who was a chemist and smart businessman who discovered palladium and rhodium at the beginning of the 19th century. He made a substantial fortune as the only person able to extract platinum from its ore. The silver coating allows the very fragile wire to be handled, and it is dissolved in nitric acid after the wire has been inserted in the glass sheath. Smaller electrodes as well as other geometries such as line electrodes can be made by lithography on coated silicon substrates.

6.6 Experimental Results Obtained with a 25-Micron UME

The data analyzed in this section were kindly obtained by Professor Frank Marken. The first measurement with the 25-micron diameter UME involved slow scanning of the potential to record a steady-state voltammogram. This raises the question of how slow the scan needs to be to maintain steady-state diffusion. If we scan very fast, the thickness of the diffusion layer will be small compared with the diameter of the electrode, and the UME will behave like a planar electrode. In Chapter 3, we looked at diffusion to a large planar electrode in response to a large amplitude potential step and found that the current decays with $\frac{1}{\sqrt{t}}$. This corresponds to an increase of the diffusion layer thickness with $\sqrt{t}$. In fact,

$$\delta(t) = \sqrt{\pi Dt} \qquad (6.22)$$

We know that to obtain steady-state hemispherical diffusion, the diffusion layer has to expand to a size that is considerably greater than the electrode diameter. If $D = 10^{-5}$ cm^2 s^{-1}, the diffusion layer thickness will be around 5 microns after 10 ms and 50 microns after 1 s. We therefore need to scan

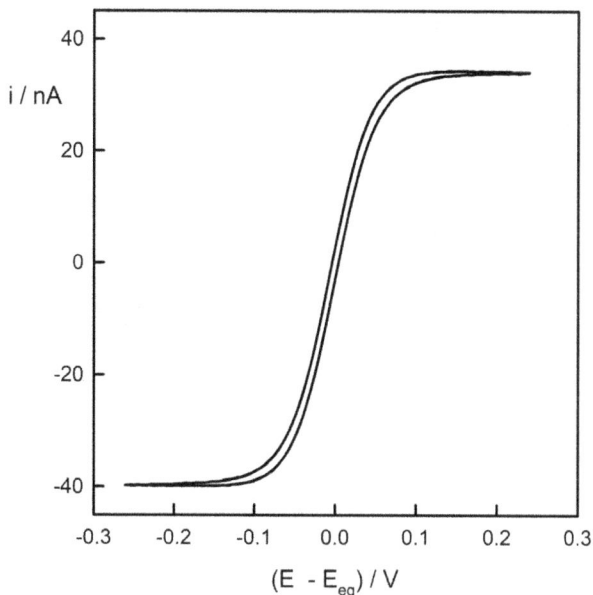

Figure 6.18. Voltammogram recorded for 25-micron diameter platinum UME in 10^{-2} M $K_4(CN)_6/10^{-2}$ M $K_3Fe(CN)_6 +$ 1.0 M KCl at a sweep rate of 5 mV s^{-1}. A platinum wire was used as reference and counter electrode. Potential vs. E_{eq}.

slowly to make sure that we maintain the steady state. Figure 6.18 shows the voltammogram recorded at a sweep rate of 5 mV s^{-1}. The small hysteresis between the forward and reverse scans shows that this sweep rate is low enough to ensure close to steady-state conditions. Note the similarity to the current potential plots for the RDE earlier in this chapter. The diffusion coefficients of the $Fe(CN)_6^{4-}$ and $Fe(CN)_6^{3-}$ ions determined from the limiting anodic and cathodic currents using equation (6.21b) are 7.0×10^{-6} cm^2 s^{-1} and 8.0×10^{-6} cm^2 s^{-1} in good agreement with the RDE results.

The similarity between the RDE and UME current potential plots suggest that perhaps we could fit the impedance of the UME with the finite Warburg that we used for the RDE. However, as we shall see, the inlaid disc geometry leads to a different response that cannot be fitted in this way. The UME impedance response has been derived by Martin Fleischmann and Stanley Pons (who are unfortunately best known for their 1989 announcement that they had observed *cold fusion* in palladium electrodes loaded electrolytically with hydrogen. This created a furore at the time, but subsequent efforts to reproduce the phenomenon have not been successful). The derivation (Fleischmann and Pons, 1988) is too long to reproduce here. The final expressions for the real and imaginary components of the impedance are given in terms of dimensionless functions Φ_4 and Φ_5 tabulated as a function of the *dimensionless frequency variable* $r_0^2 \omega / D$.

$$\text{Re}[Z_{\text{UME}}] = \frac{4RT}{\pi n^2 F^2 D^{1/2} \omega^{1/2} r_0^2 C} \Phi_4 \tag{6.23a}$$

and

$$\text{Im}[Z_{\text{UME}}] = -\frac{4RT}{\pi n^2 F^2 D^{1/2} \omega^{1/2} r_0^2 C} \Phi_5 \tag{6.23b}$$

Figure 6.19 shows plots of the functions Φ_4 and Φ_5 as a function of $r_0^2 \omega / D$ constructed using the data tabulated by Fleischmann and Pons. Note that the two functions approach each other at high frequencies, which means that the real and imaginary components of the impedance will have equal magnitude, and both will be proportional to $\omega^{-1/2}$. You should recognize this as the characteristic of the high-frequency limit of Warburg diffusion impedances when the Nyquist plot is a straight line with an angle of $-45°$ to the real axis.

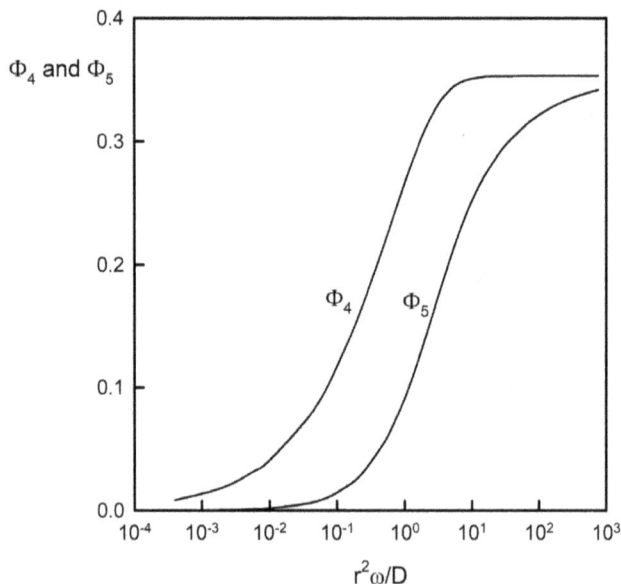

Figure 6.19. Plots of the functions Φ_4 and Φ_5 as a function of $r_0^2 \omega / D$ from the data tabulated by Fleischmann and Pons (Fleischmann and Pons, 1988).

So, what is the best way to measure the impedance of a UME? Normally, we use a potentiostat in a three-electrode configuration. As we shall see, the impedance at low frequencies is of the order of 10^6 Ω, which means of course that the ac currents flowing are very small. This can lead to substantial problems at high frequencies. The system is therefore susceptible to noise from the surroundings, and for this reason we placed the cell in an earthed metal box that acts as a *Faraday cage* (not exactly a new idea. Michael Faraday came up with it in 1836). Further improvement of the system response at very high frequencies was achieved by using a two-electrode configuration in which a platinum wire acted as both counter and

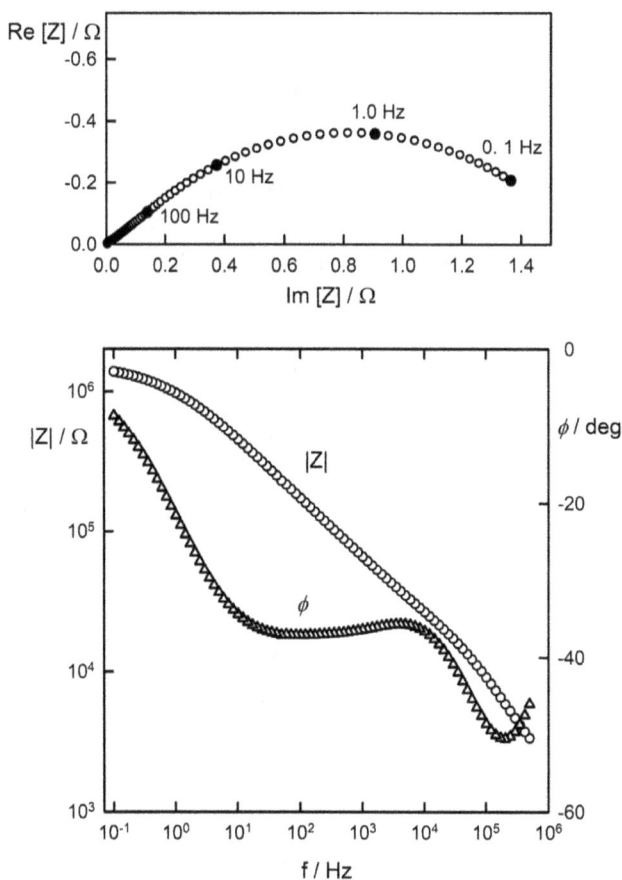

Figure 6.20. Nyquist and Bode plots of the impedance of a 25-micron UME in 10^{-2} M $K_4(CN)_6/10^{-2}$ M $K_3Fe(CN)_6+$ 1.0 M KCl.

reference electrode. Since the ac current passed is in the nA range, the platinum wire remains at the equilibrium potential. In the two-electrode mode, the measured impedance corresponds to a series connection of the impedances of the counter and working electrodes, but since the surface area of the counter electrode is many orders of magnitude higher than that of the UME, its impedance is negligible.

The impedance response of the 25-micron platinum UME for the hexacyanoferrate redox system is shown in Figure 6.20. The phase angle has an almost constant value between 10 Hz and 1 kHz, before showing a peak at higher frequencies. The peak arises from the faradaic resistance in parallel with the double-layer capacitance. The semicircle associated with this part of the modified Randles equivalent circuit is only evident if we expand the high-frequency part of the Nyquist plot.

Figure 6.21 shows a fit of the Nyquist plot to the predictions of Fleischmann and Pons. The inset is an expansion of the high-frequency region showing the partial semicircle due to the parallel combination of R_F and C_{dl}. Note that the shape of the UME impedance is not the same as that of a finite Warburg impedance for a planar electrode, so there would be no sense in trying to fit it with such an impedance in ZView®.

It is quite difficult to show the quality of a fit in a Nyquist plot because the high-frequency data are strongly compressed. A better way of showing the fit is illustrated in Figure 6.22, where the real and imaginary parts of the UME impedance are plotted vs. $\log_{10} f$. Clearly, the fit is satisfactory. For further details of the validation of the theory of Fleischmann and Pons using the hexacyanoferrate system, see (Abrantes *et al.*, 1988).

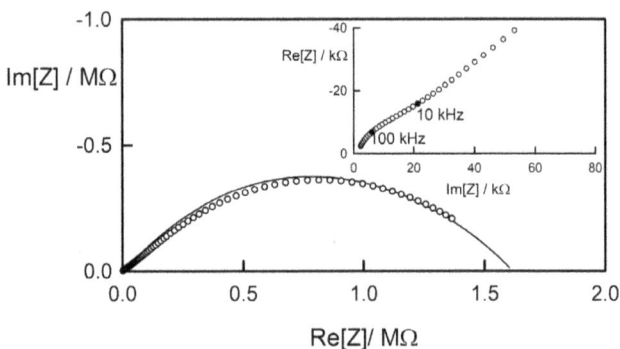

Figure 6.21. Fitting the experimental UME impedance to the theory of Fleischmann and Pons (Fleischmann and Pons, 1988). Points – experimental (see Figure 6.20). Line – fit to theory. The inset shows the partial semicircle at high frequencies arising from the parallel combination of R_F and C_{dl}.

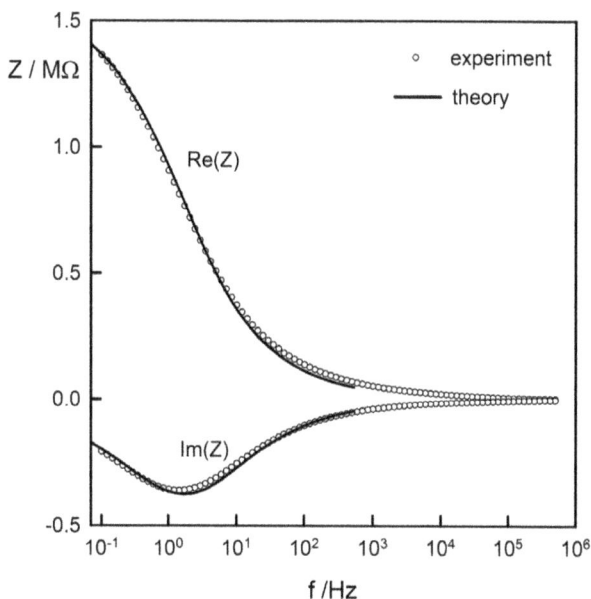

Figure 6.22. Fitting the real and imaginary parts of the UME response as a function of frequency. Points – experimental, lines – fit to theory.

It is unlikely that impedance measurements on UMEs will be important in practical applications, but it is useful to see how changing electrode geometry impacts on the formulation of the diffusion impedance. Nevertheless, UMEs are widely used for ultrafast measurements such as cyclic voltammetry because the currents are small. Using UMEs, it has proved possible to carry out cyclic voltammetry at sweep rates of more than a million volts a second. Under these conditions, the diffusion layer thickness is much smaller than the diameter of the electrodes. So, for example, if we scan at 10,000 V s^{-1} over a range of, say 500 mV, it takes only 100 μs to complete the scan. In this time the diffusion layer thickness ($\delta = \sqrt{\pi D t}$) will have grown only to 0.5 microns.

References

Abrantes, L. M., Fleischmann, M., Peter, L. M., Pons, S. & Scharifker, B. R. 1988. On the diffusional impedance of microdisc electrodes. *Journal of Electroanalytical Chemistry*, 256, 229–233.

Fleischmann, M. & Pons, S. 1988. The behavior of microdisk and microring electrodes – Mass-transport to the disk in the unsteady state – The AC response. *Journal of Electroanalytical Chemistry*, 250, 277–283.

Chapter 7

Photoelectrochemical Impedance Spectroscopy of Dye-Sensitized Solar Cells and Metal Halide Perovskite Cells

7.1 Introduction

The technique by which impedance measurements are made under steady illumination of an electrode or a solar cell is referred to as *photoelectrochemical impedance spectroscopy* (PEIS). In this chapter, we look at the application of PEIS to the study of two types of solar cell that have been the subject of intense research in recent decades. The first is the dye-sensitized solar cell (DSSC), which was developed in the 1990s by Brian O'Regan and Michael Grätzel (Oregan and Grätzel, 1991) at the Ecole Polytechnique Federale de Lausanne. The second is the metal halide perovskite solar cell (PSC), which was developed originally as an offshoot of DSSC research and has gone on to achieve remarkable efficiencies in an extraordinarily short time (at the time of writing, over 25%). The DSSC is unlike a conventional solid-state solar cell such as the familiar silicon solar cell since it is essentially a thin-layer *electrochemical* cell that can generate electrical power (current and voltage) under illumination. The key element of the cell is a thin (10–15 microns) mesoporous layer of sintered TiO_2 particles coated with a light absorbing dye (usually a ruthenium bipyridyl compound, although organic dyes have also been used successfully). The sintered TiO_2 layer is deposited on glass coated with a transparent layer of highly conducting fluorine-doped tin oxide: FTO. The porous TiO_2 layer is permeated by a non-aqueous electrolyte containing I_3^- and I^- ions, and the cell is completed by a second FTO-coated glass plate on which a layer of platinum nanoparticles has been deposited. The PSC was developed initially by Tsutomu Miyasaka and his colleagues (Kojima *et al.*, 2009) as a type of DSSC in which the sensitizer dye was replaced by the lead halide perovskites methylammonium lead bromide and iodide: MAPbBr$_3$ and MAPbI$_3$ (MA = CH$_3$NH$_3^+$). These cells proved to be unstable due to corrosion of the perovskite sensitizers, but when other researchers replaced the liquid electrolyte by an organic hole conductor, much more stable cells were obtained. Since then, PSC efficiencies rocketed upwards, although long-term stability remains a problem.

7.2   PEIS of Solar Cells: What Do We Expect to See?

We start by looking at what we expect for the impedance of an ideal solid-state solar cell. We then add a dose of reality before moving on to analyze the impedance data of the DSSC and the PSC.

A solar cell generally consists of a semiconducting *absorber* that is contacted on both sides by *selective contacts* Photons absorbed in the semiconductor generate *electron–hole pairs*, i.e., electrons are promoted from the valence band to the conduction band, leaving vacancies (holes) in the valence band. Consequently, the steady-state concentrations of electrons and holes under illumination are higher than their equilibrium (dark) values. In the dark, the equilibrium concentrations of both electrons and holes are determined by the position of *Fermi energy*, or *Fermi level*, E_F, relative to the energy of the valence and conduction bands. For an explanation of the Fermi level and quasi-Fermi levels, see Appendix 7.2 at the end of this chapter.

Under illumination, the concentrations of electrons and holes determine separate *quasi-Fermi levels* (QFLS) that move closer to the respective band energies as the concentrations increase. This *splitting of the Fermi levels* leads to the generation of a *photovoltage*. PEIS measurements are usually made under open-circuit conditions, where typically initially flat QFLs in the absorber are perturbed by a small ac voltage applied across the device. The quasi-Fermi levels for electron and holes are labelled $_nE_F$ and $_pE_F$ to distinguish them from the equilibrium Fermi level E_F. $_nE_F$ moves towards the conduction band as the electron concentration increases, and $_pE_F$ moves towards the valence band as the hole concentration increases.

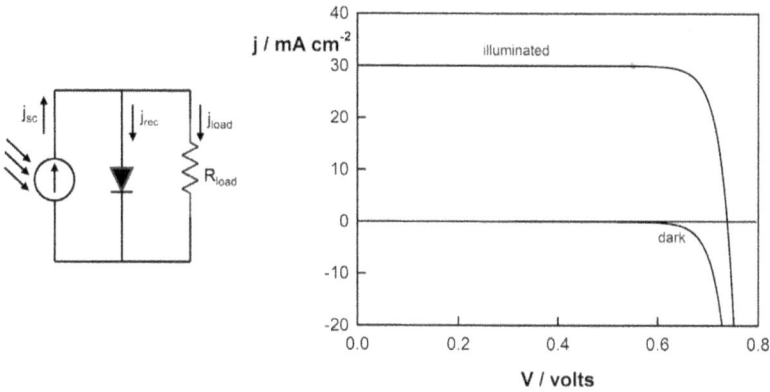

Figure 7.1. Simple equivalent circuit and current voltage plots for an ideal solar cell ($m = 1$) in the dark and under solar illumination. $j_{sc} = 30$ mA cm^{-2}, $j_s = 10^{-14}$ A cm^{-2}. The open-circuit voltage V_{oc} corresponds to the voltage at which the current under illumination passes through zero. PEIS measurements are normally made at V_{oc}.

In textbooks, you will find a simple equivalent circuit for an *ideal* solar cell like the one shown in Figure 7.1. In the dark, the cell behaves as a *diode*, passing current easily in one direction but almost blocking current flow in the other direction. Under illumination, a constant (i.e., voltage-independent) photocurrent is 'added' to the dark current to give the overall voltage-dependent photocurrent, $j(V)$, which is given by

$$j(V) = j_{sc} - j_s \left(e^{\frac{qV}{mk_BT}} - 1 \right) \qquad (7.1)$$

In photovoltaics, the photocurrent is taken to be positive and the forward dark current, negative. This the opposite sign to that normally used in electronics. Here, j_s is the *reverse saturation current density* and j_{sc} is the *short circuit current density* under illumination. j_s is associated with thermally generated electron–hole pairs. For good solar cells, j_s is many orders of magnitude smaller than the short-circuit current density under solar illumination. Equation (7.1) includes a factor m known as the *diode ideality factor*. For an ideal diode, $m = 1$. Real solar cells generally have m values greater than 1. In the dark, $j_{sc} = 0$, so the first term describes the dark jV plot. In practice, real solar cells may deviate significantly from this ideal behaviour, but we will not go into details here. All we need to note at this point is that equation (7.1) describes the current voltage plot of an 'ideal' solar cell shown in Figure 7.1. At the open-circuit potential under illumination, the net current is zero, which means that the first and second terms in equation (7.1) sum to zero. It follows that the open-circuit voltage is given by

$$V_{\text{oc}} = \frac{mk_BT}{q}\ln\left(\frac{j_{\text{sc}}}{j_s} + 1\right) \tag{7.2}$$

In practice, we can forget the 1, since $j_{\text{sc}}/j_s \gg 1$. This means that the open-circuit voltage increases logarithmically with light intensity. At the open circuit, none of the photogenerated electron–hole pairs are extracted from the solar cell, in other words the rate of generation of electron–hole pairs exactly equals the rate of their recombination. The recombination current $j_{\text{rec}} = j_{\text{sc}}$.

Now consider what happens when we add a small additional voltage to the open-circuit voltage. The only term in equation (7.1) that will be affected is the exponential. For $qV_{\text{oc}}/mk_BT \gg 1$ and a small change in voltage δV, equation (7.1) becomes

$$j + \delta j = j_{\text{sc}} - j_s e^{\frac{q(V_{\text{oc}}+\delta V)}{mk_BT}} = j_{\text{sc}} - j_s e^{\frac{qV_{\text{oc}}}{mk_BT}}\left(1 + \frac{q\delta V}{mk_BT}\right) \tag{7.3}$$

Noting that the net current is zero at open circuit, it follows that the increment in current is

$$\delta j = -j_s e^{\frac{qV_{\text{oc}}}{mk_BT}}\frac{q\delta V}{mk_BT} = -j_{\text{sc}}\frac{q\delta V}{mk_BT} = j_{\text{rec}}\frac{q\delta V}{mk_BT} \tag{7.4}$$

This allows us to obtain a *recombination resistance* defined as

$$R_{\text{rec}} = \frac{\delta V}{\delta j} = \frac{mk_BT}{qj_{\text{sc}}} = \frac{mk_BT}{qj_{\text{rec}}} = -\frac{mk_BT}{qj_{\text{sc}}} \tag{7.5}$$

As we have seen earlier, the charge associated with the electrons and holes present in an illuminated solar cell depends on the positions of the electron and hole QFLs. This allows us to define what is termed a *chemical capacitance*, C_{chem}, which is described in more detail in Appendix 7.3 at the end of this chapter. If the semiconductor has *trapping states*, we also need to consider how the electronic charge stored in traps changes with the QFL splitting, since this also contributes to the total chemical capacitance. In addition to the chemical capacitance associated with free and trapped electronic charges, we also have a *geometric capacitance* arising from the capacitor-like arrangement of the solar cell absorber between two conducting contacts. Finally, in any real solar cell we have an ohmic *series resistance* R_{series} arising from the contacts and a *shunt resistance* due to leakage pathways between the contacts. Our small amplitude equivalent circuit for open-circuit PEIS therefore looks like Figure 7.2. Note that the circuit does not contain the current source corresponding to the generation

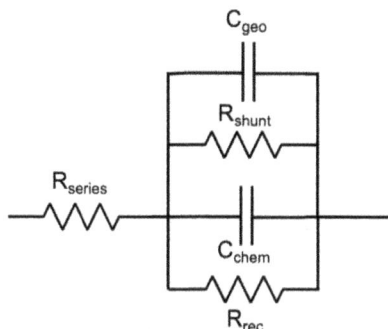

Figure 7.2. Small amplitude equivalent circuit for PEIS of an illuminated solar cell showing the chemical capacitance and recombination resistance. The circuit also includes series and shunt resistances as well as the geometric capacitance. The chemical capacitance and geometric capacitance will of course add together because they are in parallel. The shunt resistance appears in parallel with the recombination resistance, but it will only become apparent at very low light intensities where the recombination resistance is high.

of electron–hole pairs. This is because in an ideal solar cell, this current source is not affected by modulation of the voltage, i.e., j_{sc} in equation (7.1) is independent of voltage. In real solar cells, however, j_{sc} may be voltage dependent, in which case the current sources do need to be included, and analysis becomes more complicated than the one given in this section. The diode has also disappeared because it is a non-linear circuit element and we have linearized the current response for small voltage changes.

Since C_{sc} and C_{geo} appear in parallel in the circuit, the measured capacitance will be the sum of the two. This raises the question – how can we deconvolute the two capacitances? An important difference between the chemical capacitance and the geometric capacitance in Figure 7.2 is worth highlighting to answer this question. The chemical capacitance is a bulk property, so it scales linearly with the thickness of the device – the thicker the device, the higher the chemical capacitance. The geometric capacitance, on the other hand, is *inversely* proportional to the device thickness. This means that thin devices like the planar PSC (absorber thickness around 500 nm) discussed in this chapter can have chemical capacitances that are smaller than the geometric capacitance. DSSCs, by contrast, utilize much thicker absorber layers that have a high concentration of trapping states, so the chemical capacitance is much higher than the geometric capacitance.

We begin with a PEIS analysis of a DSSC, which allows us to introduce *distributed impedance elements* that are available in ZView® for modelling porous systems. We then move on to look at the impedance of PSCs, which is complicated by the fact that the metal halide perovskites are *mixed electronic/ionic conductors*.

7.3 Fabricating the DSSC

The DSSC used in this experiment was expertly fabricated in our laboratory by Hongxia Wang, who is now a professor at the Queensland University of Technology. The construction method is illustrated in Figure 7.3. The first step involves 'doctor blading' a colloidal paste (Dyesol DSL-18-N) made up of very small anatase (TiO_2) particles – typically 20–30 nm in size – onto conducting fluorine-doped tin oxide (FTO)-coated glass (TEC 15). Doctor blading involves defining an area and thickness using an adhesive tape mask and then spreading a paste into the recessed area with a sharp knife edge – here a glass microscope slide held sideways. The FTO glass was pre-coated with a very thin compact layer of TiO_2 to prevent losses due to electron transfer to I_3^- occurring via exposed FTO. The colloidal TiO_2 film was sintered at 500°C to produce a 13 micron-thick, porous layer with a very high internal surface area. The coated glass plate was then immersed in a dilute solution of a ruthenium bipyridyl sensitizer dye (Dyesol N719: full chemical name di-tetrabutylammonium cis-bis(isothiocyanato) bis(2,2'-bipyridyl-4,4'-dicarboxylato)ruthenium(II)) to produce a monolayer coating on the TiO_2. The thin layer cell was then

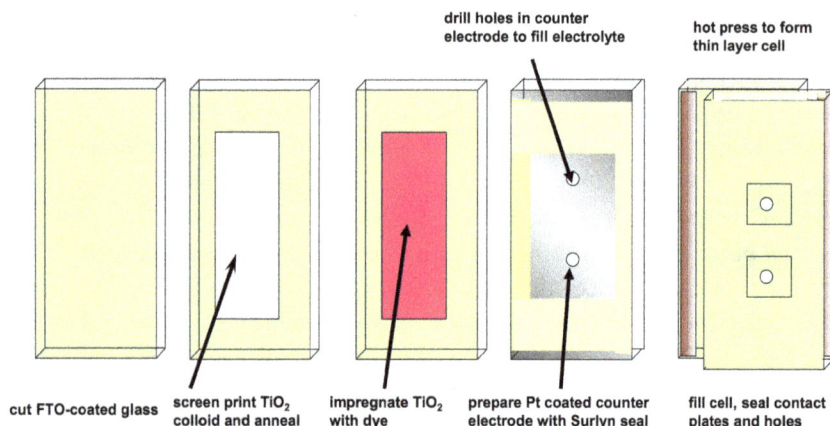

Figure 7.3. Steps in the fabrication of the DSSC used in the experiments discussed in this chapter.

made by heat-sealing on a second FTO plate coated with a nanocrystalline Pt catalyst using a thin thermoplastic Surlyn® (Solaronix) gasket. The Pt-coated plate has holes drilled to allow the injection of electrolyte and escape of air. In high performance cells, the electrolyte normally consists of LiI and I_2 in an organic solvent like acetonitrile. For high stability DSSCs, like the one studied here, the organic solvent is replaced by a room temperature ionic liquid (RTIL) to avoid losses by leakage and evaporation (the RTIL mixture used for this cell was 0.05 M I2, 0.45 M N-methylbenzimidazole, 0.1 M guanidinium thiocyanate in 1-propyl-3-methylimidazolium iodide + 1-ethyl-3-methylimidazolium thiocyanate (3:2, v/v)). The holes were heat-sealed with microscope cover slips and Surlyn®. The electrolyte-filled gap determined by the gasket is around 25 microns.

7.4 How the DSSC Works

When a molecule of the light-harvesting ruthenium sensitizer dye absorbs a photon, its excited state injects an electron from the HOMO orbital into the conduction band of the TiO_2, leaving the dye in its oxidized state. An iodide ion in the I_3^-/I^- electrolyte that permeates the pores of the layer very rapidly donates an electron to the oxidized dye, regenerating its original state ready for absorbing the next photon. This regeneration process generates I_3^- ions that can diffuse out of the porous TiO_2 layer. The injection of electrons from the dye into the TiO_2 creates a current that flows via the external circuit to the Pt-coated counter electrode, where the electrons reduce the I_3^- ions that diffuse across the narrow electrolyte gap from the TiO_2 layer. Finally, the I^- ions produced at the counter electrode diffuse back to the TiO_2 layer to complete the cyclic process, as shown in Figure 7.4.

The steps involved in the generation of current under illumination are as follows:

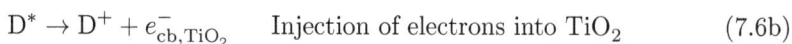

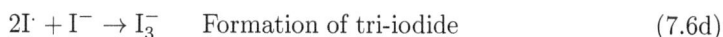

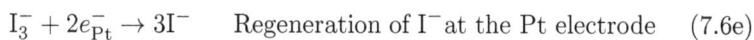

$$D + h\upsilon \rightarrow D^* \qquad \text{Photoexcitation of dye} \qquad (7.6a)$$

$$D^* \rightarrow D^+ + e_{cb,TiO_2}^- \qquad \text{Injection of electrons into } TiO_2 \qquad (7.6b)$$

$$D^+ + I^- \rightarrow D + I^{\cdot} \qquad \text{Regeneration of dye} \qquad (7.6c)$$

$$2I^{\cdot} + I^- \rightarrow I_3^- \qquad \text{Formation of tri-iodide} \qquad (7.6d)$$

$$I_3^- + 2e_{Pt}^- \rightarrow 3I^- \qquad \text{Regeneration of } I^- \text{at the Pt electrode} \qquad (7.6e)$$

$$I_3^- + 2e_{TiO_2}^- \rightarrow 3I^- \qquad \text{Back reaction of injected electrons} \qquad (7.6f)$$

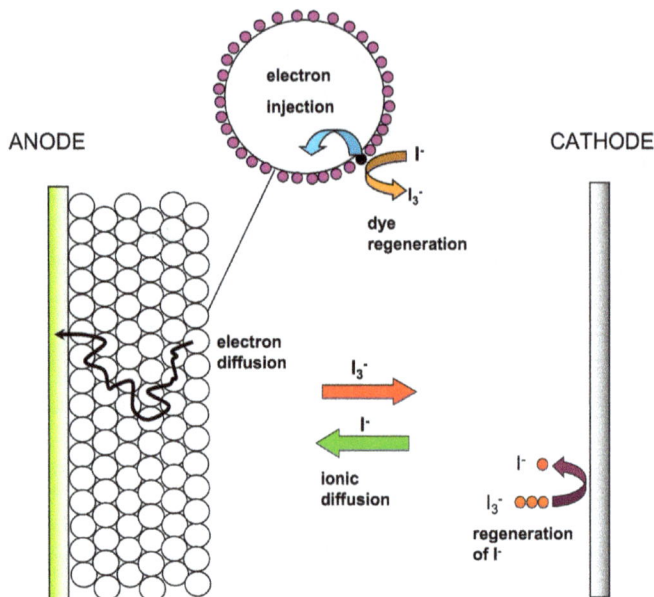

Figure 7.4. How the DSSC works. The figure illustrates the photophysical and electrochemical processes taking place in an illuminated DSSC under conditions when a current flows in the external circuit.

Equations (7.6e) and (7.6f) are important since they determine whether a DSSC works or not. We want the reduction of I_3^- at the platinum electrode to be as *fast* as possible to avoid needing a substantial overvoltage to drive the reaction because this overvoltage represents a power loss. On the other hand, we want the reduction of I_3^- by electrons in the TiO_2 to be as *slow* as possible so that the electron concentration can build up, increasing the cell voltage and reducing current losses during transport of electrons to the FTO contact. The reduction of I_3^- is fast on the platinized counter electrode because iodine atoms formed by dissociation of I_3^- ions adsorb very strongly on platinum, and electron transfer to regenerate I^- is very fast. In other words, platinum acts as an *electrocatalyst* By contrast, iodine does not adsorb on TiO_2, and reduction of I_3^- to I^- proceeds much more slowly in two electron transfer steps via the I_2^- radical anion. If the I_3^-/I^- redox couple is replaced by a fast one-electron redox couple like ferrocene/ferrocinium, the DSSC produces only tiny amounts of power because electrons injected into the TiO_2 quickly cross back to the redox system, effectively short-circuiting the cell. Typical open-circuit DSSC voltages are in the range 0.7–0.9 V

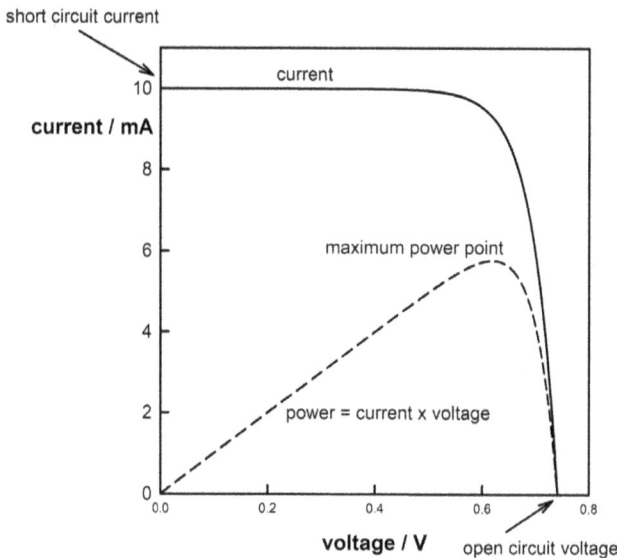

Figure 7.5.   Generalized jV characteristic of a DSSC (or any solar cell) showing the maximum power point.

under solar illumination. The voltage corresponds to the difference between $_nE_{F,TiO_2}$, the electron Fermi level in the TiO$_2$, and $E_{F,I_3^-/I^-}$, the redox Fermi level of the I_3^-/I^- couple at the platinized counter electrode.

Of course, for a solar cell to be useful it must generate *power*. The power generated by a solar cell is the product of the current and the voltage. If we connect our illuminated DSSC to an external *load resistor*, we can control the amount of current flowing by varying the resistance. For each resistance value, we will have a voltage across the resistor and a current through the resistor. The power dissipated in the resistor is the product of current × voltage. The current–voltage characteristic or *IV characteristic* of a typical DSSC is shown in Figure 7.5. The figure also shows the *maximum power point* of the cell, which is where we would like to operate the cell for maximum efficiency.

7.5 Modelling the Impedance of the DSSC: More on Distributed Elements

Solar cells are *two-terminal devices*. Normally, when we make measurements with a potentiostat in a three-electrode configuration, we measure only the impedance of the working electrode and can forget about the impedance of the counter electrode. However, in the case of a two-terminal device like a solar cell or a battery, we measure the impedance of the whole device, which in the case of a battery, for example, consists of the series connection of anode, electrolyte and cathode. Importantly, we no longer have a reference electrode. Now that we have a better idea of how the DSSC works, we can break the cell up into different parts to associate them with appropriate impedance elements. The easiest is the I_3^-/I^- electron transfer reaction at the platinized counter electrode. We can represent this by the usual parallel combination of Faradaic resistance R_F and double-layer capacitance C_{dl}. Next, we consider the diffusion of I_3^- and I^- ions across the thin-layer cell as shown in Figure 7.4. We saw in Chapter 3 that this process can be represented by a finite Warburg (short) element. So far, we are in familiar territory. But now we come to the mesoporous TiO_2 layer coated with dye and permeated by the redox electrolyte. This is obviously a porous electrode, but not one consisting of a conducting material like, say, carbon. Instead, we have a wide band gap oxide semiconductor, TiO_2. In the dark, there are essentially *no* electrons in the TiO_2 – it is an insulator. This is because the *redox Fermi level* of the I_3^-/I^- electrolyte is around 1 eV lower than the conduction band energy of the TiO_2, so that any electrons present initially due to n-type doping of the oxide are 'sucked out' by the redox couple, leaving the equilibrium electron concentration close to zero. Under illumination electrons are injected from the sensitizer dye into the oxide particles. But how is this possible? Normally, injection of electrons into an insulator generates a *space charge*, and a large electric field strongly inhibits current flow. The key to this mystery is the fact that the oxide is not a bulk solid – it is *mesoporous* (a mesoporous material is defined by IUPAC as one that has pores with diameters between 2 and 50 nm). The pores are filled with high ionic strength electrolyte, and the displacement of the ions and orientation of solvent dipoles shields the electron charge in the TiO_2 in such a way that high electron concentrations can be built up under illumination without problems. When we draw current from the solar cell, the electrons in the TiO_2 move towards the FTO substrate by *trap-controlled diffusion* because the shielding effect of the ions in the pores prevents the build-up of any directional electric field across the porous layer. At the same time, these electrons may transfer back to the electrolyte by reacting with I_3^- ions.

Trap-controlled diffusion involves capture of conduction band electrons by traps and thermal release back to the conduction band. Most electrons in the DSSC are in trap states, with only the few in the conduction band contributing to the current. The origin of the trapping states in mesoporous TiO_2 is poorly understood.

Summarizing, for the sensitized TiO_2 electrode we need an equivalent circuit that represents (i) electron injection, (ii) electron transport to the substrate, (iii) electron transfer from TiO_2 to the electrolyte. This circuit needs to be *distributed* in the space corresponding to the thickness of the mesoporous layer. In other words, we are looking for a *distributed circuit element* that represents all the processes listed above. Figure 7.6 is such an equivalent circuit. It includes the following distributed elements: (a) *current sources* representing the injection of electrons from the photoexcited dye; (b) *transport resistances* R_{tr} in the top rail to describe the movement of electrons in the mesoporous film; (c) shunting *charge transfer resistances* R_{ct} to model the back transfer of electrons from the TiO_2 to the electrolyte; (d) parallel *chemical capacitances* C_{chem} to describe the storage of electrons in the TiO_2 film; (e) resistances R_{el} in the bottom rail to model the transport of ions in the electrolyte. The time constant $\tau_n = R_{ct}C_{chem}$ defines the *effective lifetime* of electrons injected into the TiO_2. This is determined by the rate of the reaction of electrons in the TiO_2 with I_3^- ions in the electrolyte.

Figure 7.6 immediately raises an important question – are the values of all the elements of the distributed circuit the same throughout the film? The answer is – only under certain circumstances. Consider first the injection of electrons represented by the current sources. The local injection current depends on the light intensity, which generally varies with distance according to Beer Lambert law.

$$I(x) = I_0 e^{-\alpha(\lambda)x} \tag{7.7a}$$

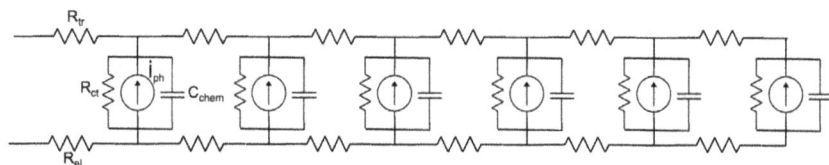

Figure 7.6. An equivalent circuit that includes: (a) distributed *current sources* i_{ph} to represent electron injection from the photoexcited dye; (b) distributed *resistors* R_{tr} to represent the transport of electrons in the mesoporous TiO_2; (c) distributed *resistors* R_{ct} to represent the back transfer of injected electrons to I_3^- ions; (d) distributed *chemical capacitances* C_{chem} to represent the storage of electrons in the TiO_2 nanoparticles; (e) distributed resistors R_{el} to describe the transport of current-carrying ions in the electrolyte. The current sources can be omitted from the PEIS analysis of the DSSC since the electron injection rate is independent of voltage.

where $\alpha(\lambda)$ is the wavelength-dependent absorption coefficient of the sensitizer dye (cm^{-1}). This corresponds to a local electron injection rate $G(x)$ given by

$$G(x) = \alpha I_0 e^{-\alpha(\lambda)x} \qquad (7.7b)$$

To ensure approximately homogeneous light absorption, we can choose an illumination wavelength at which the dye absorbs weakly. This is at the long wavelength onset of the dye absorption spectrum. What about the resistors and capacitors? It turns out that if we make impedance measurements with weakly absorbed light at the open-circuit potential, the electron concentration is nearly uniform and the ac currents in the TiO$_2$ and in the electrolyte are small, so that we can consider the values of the distributed R_{trans} and C_{chem} to be independent of distance. At high light intensities, the concentrations of I$_3^-$ and I$^-$ will vary with distance across the cell under short-circuit conditions, but at open circuit where no current flows, their concentrations, and hence the R_{ct} values, will also be uniform.

A second question then arises. What is the *physical meaning* of the transport resistance and the chemical capacitance? We are used to the idea that the presence of a resistor implies the presence of a voltage. Similarly, we associate capacitors with a voltage difference between two conducting plates. Here, we run into a problem. The shielding of electronic charge by the electrolyte in the mesoporous TiO$_2$ layer means that there is no directional electrical field to move electrons to the FTO contact. Instead, they move by *diffusion* down a concentration gradient. Similarly, charge storage occurs in the *bulk* of the mesoporous layer without the build-up of an electrostatic potential difference. As mentioned earlier, this means that the chemical capacitance is proportional to the layer thickness, whereas the geometric capacitance depends *inversely* on the thickness of the dielectric between the capacitor plates. To avoid digression, a more detailed discussion of the chemical capacitance and transport resistance is given in Appendix 7.1 at the end of this chapter. All we need to note here is that there is a self-consistent definition of all the elements in our equivalent circuit. We can stop worrying.

The final question is – how is the performance of the DSSC related to the values of the circuit elements in Figure 7.6? A key parameter in this context is the *electron diffusion length L_n*. This corresponds to the average distance that an electron diffuses through the mesoporous TiO$_2$ before it is lost by transfer to I$_3^-$. An efficient DSSC will have an electron diffusion length that is considerably longer than the thickness of the TiO$_2$

layer so that almost all injected electrons are extracted at the FTO contact. The competition between electron transport and electron transfer to I_3^- is expressed by the relative values of R_{tr} and R_{ct}, and it can be shown that the electron diffusion length, L_n, is given by

$$L_n = d\sqrt{\frac{R_{ct}}{R_{tr}}} \tag{7.8}$$

where d is the thickness of the TiO$_2$ layer. Impedance measurements under illumination therefore offer a simple way of measuring the electron diffusion length in a DSSC if we have a suitable distributed circuit to model the EIS response.

We have already met distributed element circuits. The de Levie pore model in Chapter 3 is an example. Generally, this type of element can arise when the impedance is associated with a finite three-dimensional region of space. For example, the different types of diffusion impedance are associated with a diffusion layer. ZView® lists a large number of different extended elements, and it is not possible to discuss them all here. A DE (distributed element) requires five parameters to be specified. A DX (extended distributed element) type requires more than five parameters to be specified. In most cases, the ZView® help file on particular DE or DX element points to the relevant literature. Here, we will look at one of these elements that is labelled DXType12: Bisquert#3. These impedance elements are based on the extensive work of Juan Bisquert, Professor of Applied Physics at the Jaume I University in Castellón, Spain. References to his papers are given in the ZView® help file. This type of distributed element has been widely used to fit the impedance of conducting polymers, electrochromic layers and dye-sensitized solar cells, and Figure 7.7 matches the circuit shown in Figure 7.6 except that it does not include the distributed current sources. In fact, we do not need to include these when modelling the impedance response because the modulation of the voltage does not affect the injection current: this is controlled by the illumination,

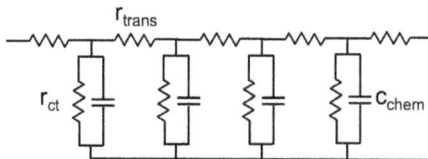

Figure 7.7. ZView® Bisquert#3 distributed element model used to analyze the impedance of the dye-sensitized solar cell under illumination at open circuit.

which does not change with time. In a good DSSC, nearly 100% of the injected electrons pass to the external circuit under short-circuit conditions. By contrast, when the solar cell is illuminated under *open-circuit* conditions, no current can pass, so the electron concentration in the TiO_2 builds up until the rate of electron injection is exactly balanced by the transfer of electrons the other way to I_3^- ions. The resulting build-up of electrons in the TiO_2 generates the open-circuit voltage. The slower the back transfer of electrons to I_3^- ions, the higher the photovoltage. In our impedance measurements of dye-sensitized solar cells, we set the dc potential to equal the open-circuit voltage, V_{oc} (typically around 0.8 V for solar intensities). This is done by setting the potentiostat to hold the dc potential at 0 V vs. reference, with the reference and secondary electrode cables both connected to one side of the solar cell and the working electrode cable connected to the other side (the two-electrode mode). The potential is then perturbed from V_{oc} by a small amplitude ac voltage signal of varying frequency. This ac voltage changes the electron concentration at the FTO contact in much the same way as it alters the concentration of redox species at the surface of an electrode.

As we saw in Figure 7.6, the top line of resistors represents the electrons moving in the TiO_2, and the bottom line corresponds to ions moving in the electrolyte that permeates the pores. For simplicity, we have assumed that the conductivity of the electrolyte is sufficiently high that we can ignore resistive losses, so the bottom line of resistors is replaced by a continuous wire.

To use the Bisquert#3 element in our modelling, we choose *DX – Extended Element* from the element type menu in the equivalent circuits window. An element labelled DX1 then appears in our circuit. Clicking on the *Value* window reveals a long list of elements (31 in my current version of ZView®. We then choose 11: *Bisquert#3* and are confronted with the list of 10 initially mysterious parameters shown in Figure 7.8.

At this point, we need to consult the help file to find out what the Bisquert#3 circuit is and what the different parameters represent. We find that the circuit that is modelled is more general than the one in Figure 7.6. The transport resistance in our circuit is replaced by a parallel combination of a resistor and a constant phase-shift element, and the chemical capacitance is also replaced by a constant phase element (CPE). There is also an additional impedance X2 representing the boundary condition for electrons at the end of the TiO_2 layer furthest from the FTO substrate. In our case, this element is absent, since electrons can only exit

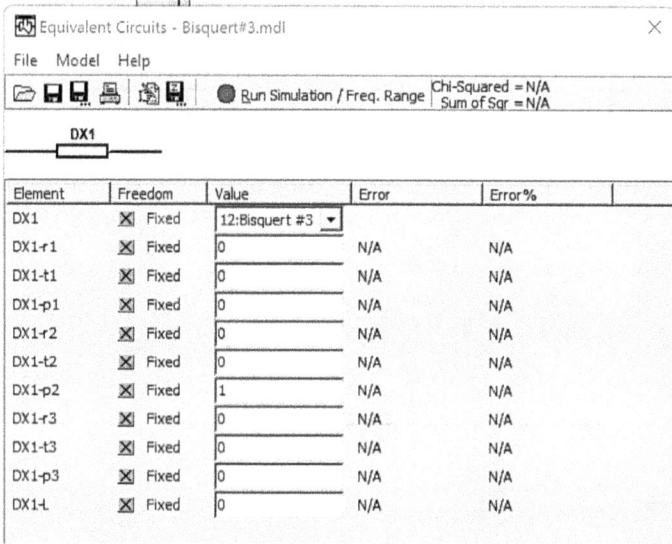

Figure 7.8. ZView® equivalent circuit screen for the Bisquert#3 distributed impedance element.

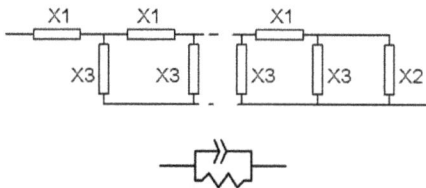

Figure 7.9. Circuit shown in the ZView® help file for Bisquert#3 distributed element. The impedances labelled X are parallel resistance/CPE circuits. Note that in the modelling of the circuit in Figure 7.7, X2 is absent (parameter values set to zero). The CPE in X3 is replaced by a capacitor (i.e., by a CPE with p fixed to equal 1). See also Table 7.1.

via the conducting glass at the bottom of the layer, we need to enter zeros in the appropriate parameter fields.

Figure 7.9 is the generalized Bisquert#3 circuit as shown in the help file.

It is not immediately clear what parameter values we should enter into the Bisquert#3 model. Luckily, further down the help page we find the information needed to relate the different elements listed in Figure 7.8 to the impedances in Figure 7.9. Table 7.1 identifies the elements and shows the values used in the modelling.

Table 7.1. Table of parameter values for the Bisquert#3 model and values used in the simulation of the impedance. An asterisk indicates that the element is absent, so the value is set to zero.

ZView® window	Element	Corresponds to	Value used
DX-R	DX-r1	R_{tr}	100 Ω
DX-T	DX-t1	CPE in X1	0*
DX-P	DX-p1	CPE in X1	0*
DX-U	DX-r2	R in X2	0*
DX-A	DX-t2	CPE in X2	0*
DX-B	DX-p2	CPE in X2	0*
DX-C	DX-r3	R_F	1000 Ω
DX-D	DX-t3	CPE in X3	10^{-4}
DX-E	DX-p3	C_{chem}	1 (capacitor)
DX-F	DX-L	Length	1 (unit length)

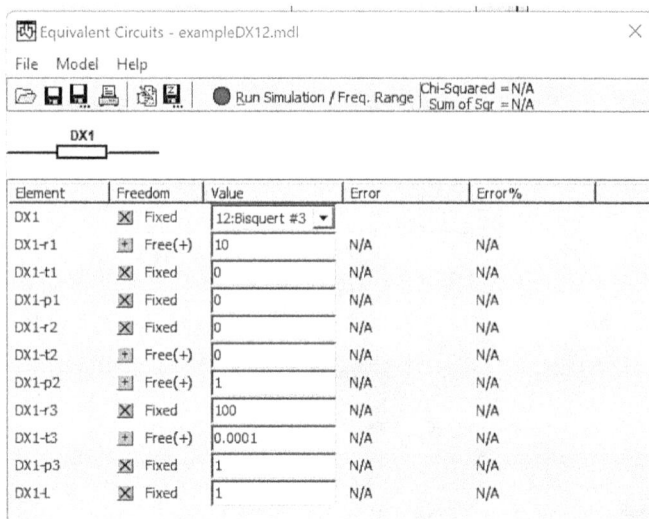

Figure 7.10. ZView® Equivalent circuit window with model Bisquert#3 circuit loaded.

We can also load the example model file which after installation of ZView® is available on your C drive at **C:\SAI\ZModels\ exampleDX12.mdl** listed in the help section. As shown in Figure 7.10, the parameter window identifies the elements as shown in Table 7.1 (the default values have been replaced by those in Table 7.1).

When the simulation is run with the values shown above, we obtain the impedance plots shown in Figure 7.11.

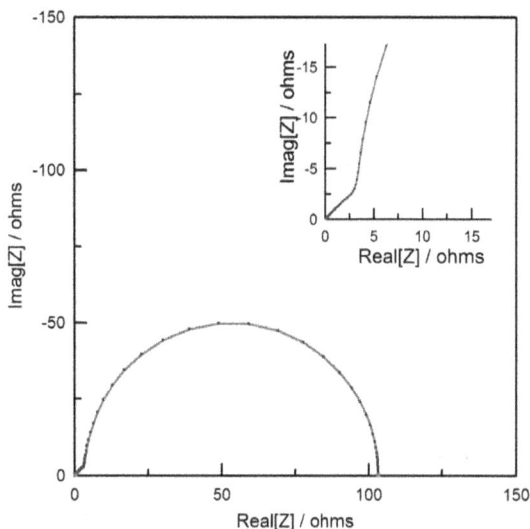

Figure 7.11. Nyquist impedance plots for the Bisquert#3 circuit (Figure 3.15) simulated in ZView® with the values shown in Figure 7.6. The insert shows the typical −45° region of the plot at high frequencies.

In this simulation example, the transport resistance per unit length (DX1-r1) is set to be 10 times smaller than the Faradaic resistance for the transfer of electrons to I_3^- ions (DX1-r3). This corresponds physically to the situation where virtually all electrons reach the back contact without being lost by transfer to I_3^- ions since $L_n/d = \sqrt{10} = 3.26$. At high frequencies, we see typical transmission line behaviour – the Nyquist plot is a straight line with a slope of −45° as shown in the expanded portion of the Nyquist plot. At lower frequencies, we see a semicircle in the Nyquist plot corresponding to the parallel combination of R_{ct} and C_{chem}. The maximum radial frequency of the semicircle corresponds to the inverse of the effective electron lifetime, τ_n, which is determined by the rate of loss of electrons by reaction with I_3^-.

So far, we have only modelled the thin mesoporous layer in the dye-sensitized solar cell. We also need to think about the diffusion of I^- and I_3^- ions across the narrow gap. We have come across this before for the *thin-layer cell*. It is modelled by the *finite length Warburg (short)*. Next, we need to consider the electron transfer reaction occurring at the Pt-coated counter electrode. We can represent this by the usual parallel combination of double-layer capacitance and Faradaic resistance. Finally, we must add the series resistance arising from the FTO (typically some tens of ohms for a 1 cm^2 cell). Our final equivalent circuit of the dye-sensitized solar cell is shown in Figure 7.12.

Now for the simulation of this circuit. The orders of magnitude of the input values shown in Figure 7.13 correspond to those seen experimentally. The result of the simulation is shown in Figure 7.14.

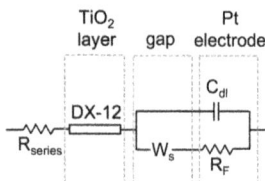

Figure 7.12. Equivalent circuit for a dye-sensitized solar cell with components representing the processes taking place in an illuminated dye-sensitized solar cell. DX1:12 is the distributed element listed as Bisquert#3 in the ZView® equivalent circuit program.

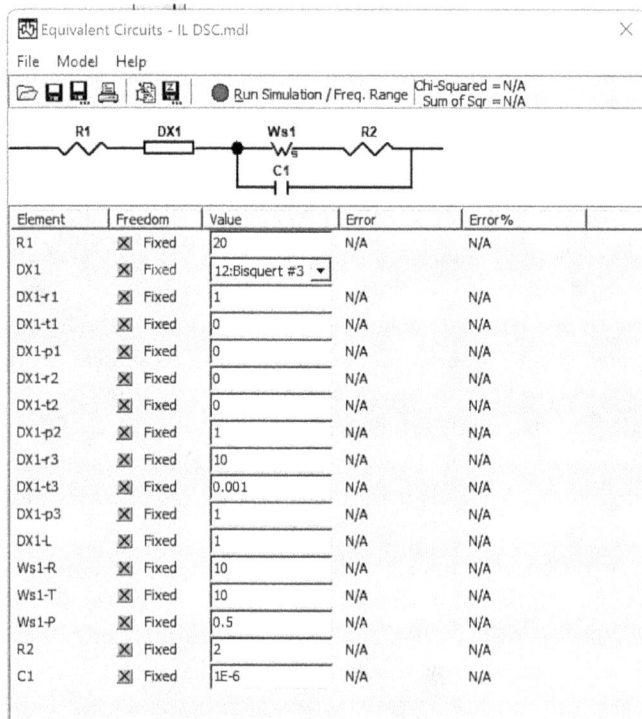

Figure 7.13. Input values used in the ZView® simulation of the equivalent circuit of a dye-sensitized solar cell. See Figure 7.14 for the result of the simulation.

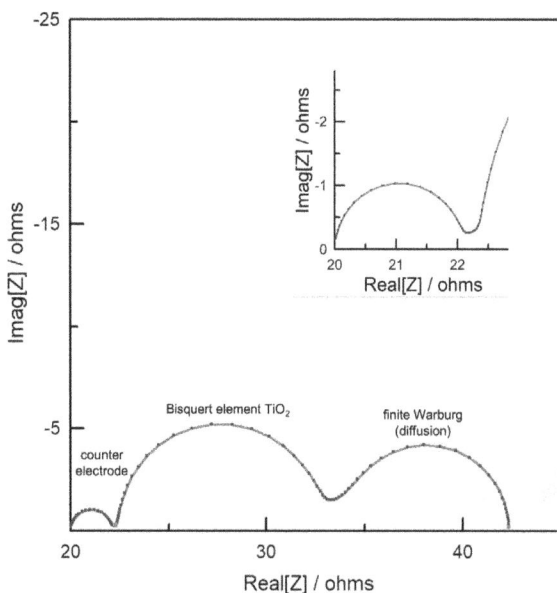

Figure 7.14. Simulated response for the circuit in Figure 7.12 with the values in Figure 7.13.

Three well-separated arcs can be seen in the Nyquist plot. The first high-frequency semicircle arises from the parallel combination of R_F and C_{dl} corresponding to the impedance of the platinized counter electrode. The next semicircle is related to the back reaction of electrons with I_3^- ions, i.e., the parallel combination of R_{ct} and C_{chem}. The high-frequency transmission line response merges into the counter electrode semicircle, but it can just be seen in the blow-up. Finally, the arc at very low frequencies is due to the finite Warburg used to represent the diffusion of ions across the electrolyte gap.

It is now time to look at the experimental result for the dye-sensitized solar cell fabricated using a room temperature ionic liquid (RTIL) electrolyte (Figure 7.15). At very low frequencies, the Nyquist plot reveals a clear Warburg response for the diffusion of ions because the RTIL is very viscous (this can be a problem for cell performance at high intensities, where narrower gaps must be used to prevent diffusion from limiting the current). At frequencies between 1 and 100 Hz, the response arises from the parallel combination of R_{ct} and C_{chem}. Finally, at high frequencies, the $-45°$ transmission line part of the Bisquert#3 response merges with the

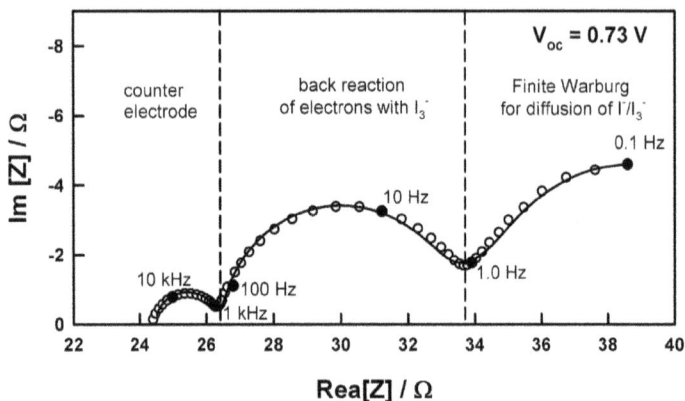

Figure 7.15. Nyquist plot of the impedance response of a 1 cm^2 dye-sensitized solar cell using an ionic liquid solvent illuminated by a solar simulator at an intensity of 1 sun. Points – experimental. Line – fit to model in Figure 7.12. Note the high viscosity of the ionic liquid gives a significant diffusion impedance. This impedance is much smaller for cells using low viscosity solvents like acetonitrile. The impedance response clearly shows the series resistance (around 24 Ω), the semicircle for the counter electrode impedance (R_F around 2 Ω), the semicircle associated with the parallel combination of Faradaic resistance and chemical capacitance associated with the back transfer of electrons to tri-iodide ions and finally the finite Warburg diffusion impedance due to diffusion of ions across the gap. Further details are given in (Wang and Peter, 2009).

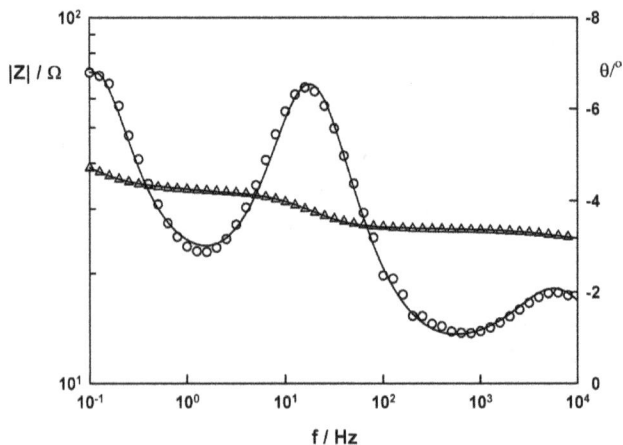

Figure 7.16. Bode plots for the impedance of the RTIL dye-sensitized solar cell. Circles – experimental $|Z|$ values. Triangles – experimental phase angle. Lines – fit to model.

RC semicircle associated with the counter electrode and is very difficult to see in the Nyquist plot. As we shall see, this leads to some uncertainty in obtaining the value of the transport resistance, R_{tr}.

The corresponding Bode plots along with the fit to the model are shown in Figure 7.16.

The values found by the ZView® fitting program are illustrated in Figure 7.17.

R1 (24.5 Ω) corresponds to the series resistance of the FTO. It has been fixed in the fitting. r1 (1.15 Ω) is the transport resistance (note the high error estimate due to the merging of the transmission line part of the response with the counter electrode impedance), and r3 (6.05 Ω)

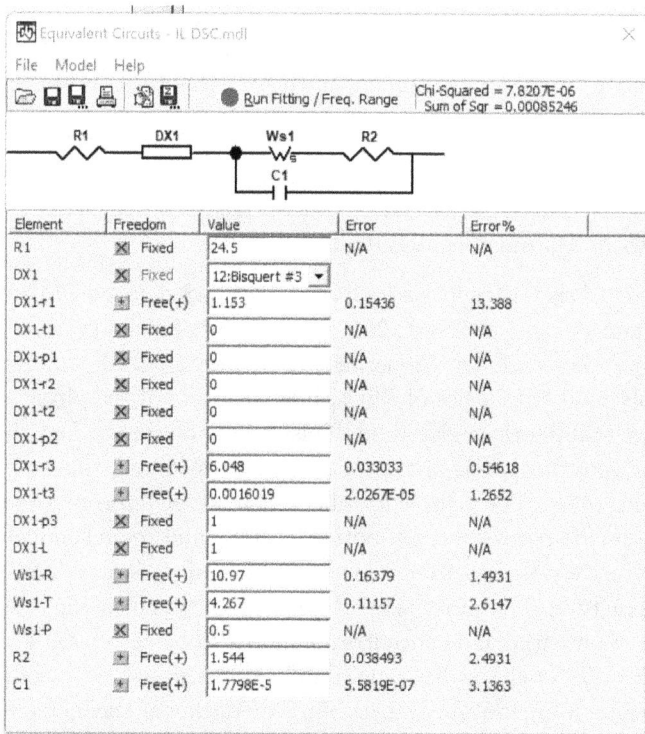

Element	Freedom	Value	Error	Error %
R1	☒ Fixed	24.5	N/A	N/A
DX1	☒ Fixed	12:Bisquert #3 ▾		
DX1-r1	⊞ Free(+)	1.153	0.15436	13.388
DX1-t1	☒ Fixed	0	N/A	N/A
DX1-p1	☒ Fixed	0	N/A	N/A
DX1-r2	☒ Fixed	0	N/A	N/A
DX1-t2	☒ Fixed	0	N/A	N/A
DX1-p2	☒ Fixed	0	N/A	N/A
DX1-r3	⊞ Free(+)	6.048	0.033033	0.54618
DX1-t3	⊞ Free(+)	0.0016019	2.0267E-05	1.2652
DX1-p3	☒ Fixed	1	N/A	N/A
DX1-L	☒ Fixed	1	N/A	N/A
Ws1-R	⊞ Free(+)	10.97	0.16379	1.4931
Ws1-T	⊞ Free(+)	4.267	0.11157	2.6147
Ws1-P	☒ Fixed	0.5	N/A	N/A
R2	⊞ Free(+)	1.544	0.038493	2.4931
C1	⊞ Free(+)	1.7798E-5	5.5819E-07	3.1363

Figure 7.17. Values of the impedance elements in the model circuit IL DSC found by the ZView® nonlinear least squares Equivalent Circuits fitting program. Note that p3 and Ws1-P are fixed. The quality of the fit is indicated by the values of the % error. The largest uncertainty is associated with the transport resistance, r1. This is because the high-frequency (transmission line) part of the Bisquert#3 impedance merges with the counter electrode semicircle.

is the Faradaic resistance for the transfer of electrons to triiodide. t3 (1.60 mF) is the chemical capacitance, Ws1-R (11.0 Ω) is the steady-state Warburg resistance, Ws-T (4.27 s) is the Warburg diffusion time constant, R2 (1.54 Ω) is the Faradaic resistance of the counter electrode and finally C1 (17.8 μF) is the double-layer capacitance of the counter electrode. We can now work out the ratio $L_n/d = \sqrt{R_{ct}/R_{trans}}$. Taking into account the fitting errors listed in the ZView® output, we find that $L_n/d = 2.31 \pm 0.16$. This means that the electron collection efficiency is very high for this DSSC.

The time constant for diffusion in this cell (4.27 s) is remarkably long, reflecting the high viscosity of the RTIL. The gap determined by the gasket is 25 microns, and since the diffusion time constant for a thin-layer cell is given by $l_{gap}^2/4D$, the mean diffusion coefficient of I_3^-/I^- ions in the RTIL is around 4×10^{-7} cm^2 s^{-1}.

To complete the analysis of this system, the *Kramers–Kronig transformation* was used to test the impedance data – this is described at the end of Chapter 4. The KK test was passed, showing that the impedance data are reliable.

7.6 PEIS of Planar Perovskite Solar Cells

In this part of the chapter, we look at some PEIS data on PSCs obtained by Dr Adam Pockett (Pockett, 2017; Pockett *et al.*, 2015) during his PhD studies in Professor Petra Cameron's laboratory in Bath. In spite of the remarkably high efficiencies of PSCs, many aspects of the physics of these devices are still poorly understood. This is reflected in the fact that many conflicting interpretations of the PEIS of perovskite cells have been given over recent years. The reason for the problems of interpretation is that the metal halide perovskites are both electronic and *ionic* conductors. The ionic conductivity arises from mobile anion and cation vacancies in the crystal structure. The movement of these ionic vacancies in response to gradients of potential and concentration needs to be accounted for because the distribution of charged vacancies influences the behaviour of electrons and holes in the material. The mobility of the ionic vacancies are many orders of magnitude smaller than the mobilities of electrons and holes, and this leads to characteristic low-frequency behaviour in the PEIS responses of perovskite cells. The interaction between the ionic and electronic systems is complex, and we will not go into this in depth here. Instead, we will contrast the PEIS response of perovskite cells with what we would expect for an ideal solar cell and then give a qualitative discussion of the reasons

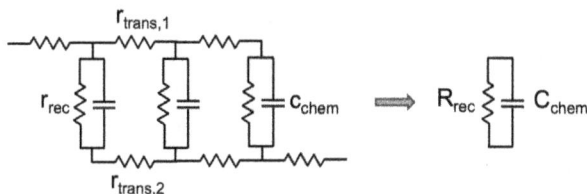

Figure 7.18.   The Bisquert#3 distributed equivalent circuit for an ideal absorber in a solar cell. If the transport resistances are small due to high carrier mobility, the circuit simplifies to the parallel RC combination shown.

behind the differences. For a comprehensive review that illustrates the confusion surrounding interpretation of PEIS data for perovskite solar cells, see Guerrero *et al.* (2021).

Let us suppose we have a solar cell with an absorber in which electron–hole pairs created by light absorption are separated using selective contacts. The electron and hole currents are determined by the gradients of the respective quasi-Fermi levels. For semiconductors with high electron and hole mobilities, only small gradients of the QFLs are needed to give the currents generated by solar illumination. However, in materials with low electron and hole mobilities, larger gradients are required. We have already seen a general equivalent circuit that describes the generation and collection of charge carriers – it is the Bisquert#3 circuit, which is shown in Figure 7.18 with transport resistances for both holes and electrons. If the transport resistances are very small (i.e., if electron and hole mobilities are high), the extended circuit collapses to a simple parallel RC circuit.

It follows that the simplest PEIS response for a solar cell would be a single semicircle in the Nyquist plot that allows us to determine R_{rec} and C_{chem}. However, we have neglected the fact that any solar cell has an ohmic series resistance and a geometric capacitance. Furthermore, there is likely to be a shunt resistance representing leakage pathways across the device as shown previously in Figure 7.2. Nevertheless, we still expect a single semicircle in the Nyquist plot. *This is not what we see for perovskite cells.* The equivalent circuit needs to take into account two additional factors. The first is the fact that charged vacancies can also move in the perovskite film, which behaves as an *ionic conductor*. The sluggish response of the low mobility ionic charges to changes in applied voltage or illumination makes establishment of a true stationary state difficult, so the current voltage plots of illuminated perovskite cells exhibit substantial hysteresis. The ionic vacancy states can build up near the contacts to form electrical double layers that compensate the electrical field that is initially formed when the insulating perovskite is contacted on two sides by materials with different work functions (TiO_2 and spiro-MeOTAD). The second factor that we need to consider is that the movement of ionic species alters the local recombination kinetics, so that recombination can no longer be represented by a simple resistor in a small signal equivalent circuit. Instead, we replace the recombination resistance by a recombination *impedance*. Figure 7.19 contrasts the equivalent circuit that we use for a 'classical' thin-film solar cell with the kind of circuit that we need to describe a planar perovskite solar cell. The additional ionic branch consisting of a series connection of resistor and capacitor of the impedance represents the charging and discharging of the ionic double layers via the series ionic resistance. The interaction between the electronic and ionic subsystems cannot be represented by linear circuit elements, so the link between the two systems is indicated by the arrow on the circuit diagram.

The perovskite solar cells used in this study consisted of a ca. 425 nm polycrystalline layer of $MAPbI_{0.75}Cl_{0.25}$ sandwiched between a spin-coated TiO_2 electron-selective layer on conducting glass and a spin-coated spiroMeOTAD   (2,2',7,7'-Tetrakis[N,N-di(4-methoxyphenyl)amino]-9,9'-spirobifluorene) hole-selective layer. An evaporated gold contact on the

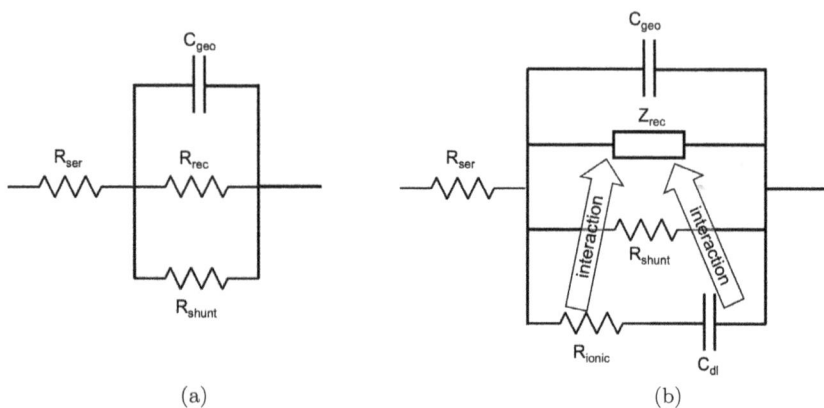

Figure 7.19. Comparison of small signal equivalent circuits for a conventional solar cell (a) with the circuit for a planar perovskite solar cell (b). Note the addition of an 'ionic branch' to represent the charging and discharging of ionic double layers at the contacts by diffusion/migration of vacancy species. Although the ionic branch will not be detected directly in PEIS measurements because the parallel impedance is too high, the movement of vacancies manifests itself via modification of the recombination impedance.

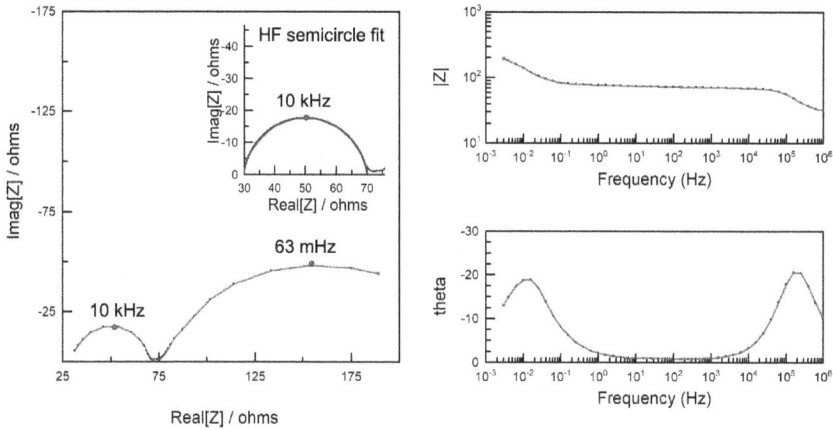

Figure 7.20. Typical Nyquist plot of the PEIS response of a perovskite solar cell measured at open circuit under illumination (425 nm). The high-frequency semicircle is what we expect for a normal solar cell. The additional response at very low frequencies is unexpected. It arises from the complex interaction between electrons and holes and mobile ionic vacancies as indicated in Figure 7.19.

spiroMeOTAD completed the cell. The cells were illuminated through the glass side. The electrode area was 0.12 cm^2. Since perovskite cells degrade in moist air, the cells were characterized in a temperature-controlled dry nitrogen environment to ensure there was no slow drift in properties.

The first surprise is that the Nyquist PEIS plots show *two* semicircles. Figure 7.20 shows that a high-frequency semicircle is observed as expected for a parallel combination of the recombination resistance and the sum of geometric and chemical capacitances. However, there is an additional very slow response that cannot be fitted very well, even using a parallel circuit of a resistor and a constant phase-shift element. There is also a strong hint of something unusual in the region between the two main responses. Clearly, the impedance behaviour of the PSC is far from what we expect for an ideal solar cell.

The high-frequency semicircle is easily identified as being the one that we expect for a solar cell. At the beginning of this chapter, we showed that the recombination resistance at the open-circuit potential is expected to be inversely proportional to the short-circuit current density.

$$R_{\text{rec}} = \frac{\delta V}{\delta j} = \frac{mk_BT}{qj_{\text{sc}}} \qquad (7.9)$$

Since we expect the short-circuit current density to vary linearly with light intensity, I_0, R_{rec} should be proportional to $1/I_0$. The variation of the parallel high-frequency capacitance with intensity should indicate whether the geometric or chemical capacitance dominates. Since the cells are thin (425 nm), C_{geo} is likely to be larger than C_{chem}. The PEIS response was therefore measured over two orders of magnitude of the light intensity (425 nm LED) using a Solartron Modulab XM system. Figure 7.21 is a log–log plot of the parallel resistance vs. intensity. The slope is close to -1, indicating that we are dealing here with the recombination resistance. By contrast, the parallel capacitance is almost constant at 25–30 nF, increasing by only 30% over two orders of magnitude of intensity. Since the relative permittivity of the perovskite is believed to be in the range 20–25, the geometric capacitance calculated from $C_{geo} = A\varepsilon\varepsilon_0/d$ should be around 5–6 nF if the cell behaves as a parallel plate capacitator. In fact, the parallel capacitance is about 4 times higher than this. The reason is not clear, but it is possible that the spin-coated spiro-MeOTAD penetrates between the perovskite crystallites, increasing the contact area and reducing the distance between the two selective contacts. The small increase in capacitance is not consistent with a significant contribution from the chemical capacitance.

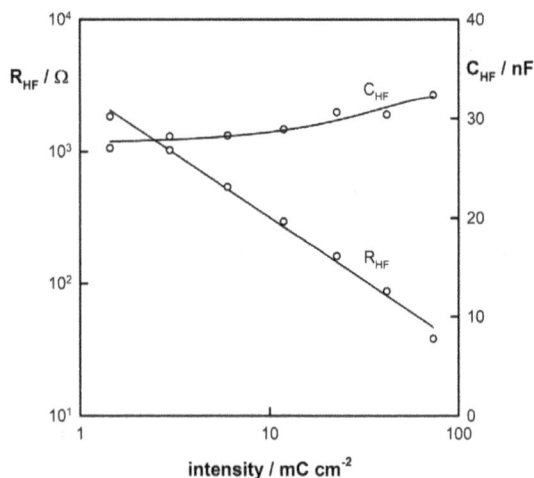

Figure 7.21. Double logarithmic plot of the high-frequency parallel resistance R_{HF} as a function of illumination intensity (425 nm LED). The slope of the log–log plot is -0.96, close to the value of -1 that we expect for an ideal solar cell. Over the same intensity range, the high-frequency parallel capacitance increases by only 30%, indicating that the geometric capacitance is dominant and the contribution from the chemical capacitance is negligible.

Turning now to the low-frequency arc that we see in the Nyquist plot, we see that it must be associated with a very slow process because the radial frequency of the maximum is in the range 40–100 mHz, corresponding to a relaxation time constant between 10 and 25 s. This relaxation time constant varies only weakly with intensity as can be seen from the Bode plot of the magnitude of the imaginary component of the impedance as a function of frequency shown in Figure 7.22. The strong shift of the high-frequency peak in Imaginary$[Z]$ to higher frequencies with increasing light intensities contrasts with the much smaller shift of the low-frequency peak.

Since the flattened shape of the low-frequency arc is difficult to fit with an equivalent circuit, even if a parallel resistance/constant phase-shift element is used, approximate values of R_{LF}, C_{LF} and $\omega_{max,LF}$ were obtained by the 'fit semicircle' routine in ZView®. This fits a *depressed semicircle* to the data, i.e., a circle with its centre below the real axis. When the values obtained by this rather crude method are plotted as a function of intensity, we see that R_{LF} decreases with light intensity in much the same way as R_{HF}. The surprising feature is the fact that C_{LF} increases with intensity in such a way that the product $R_{LF}C_{LF}$ *remains almost constant*. This raises

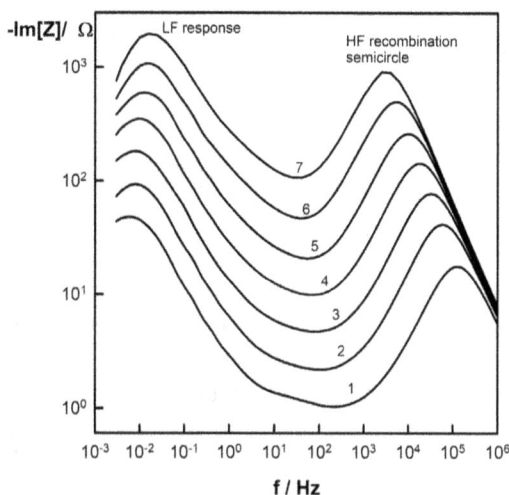

Figure 7.22. Bode plot of the magnitude of the imaginary component of the PSC impedance. The frequencies corresponding to the maximum in the low- and high-frequency arcs can be clearly seen. Note that the high-frequency peak shifts to higher frequencies as the intensity is increased. This is because the recombination resistance is inversely proportional to light intensity. The low-frequency peak, on the other hand, shifts very little, indicating that whatever process is responsible for the low-frequency response, it is only weakly intensity dependent.

an important point – are the values of R_{LF} and C_{LF} meaningful on their own, or is it only their product, i.e., the time constant $R_{LF}C_{LF}$ that has physical significance? If C_{LF} is really a capacitance, how can we explain the extraordinarily high values – greater than 1 F cm^{-2} at the highest intensity? Initially, it was suggested that this high value corresponded to a *giant dielectric effect* (Juarez-Perez *et al.*, 2014). If we take the ratio C_{LF}/C_{HF} at the highest intensity, it is 4.6×10^6!

If the static dielectric constant at higher frequencies is around 20, this would mean that the low-frequency dielectric constant is nearly 10^8. This is physically implausible, so we must look for other explanations.

The key point is that there is some slow relaxation process that is influencing the PEIS response. The most obvious explanation is that not only are electrons and holes moving in the cell, but also ionic vacancies. These vacancies can move to the interfaces with the contacts to form electrical double layers like those present in a metal in a normal electrolyte. The slow charging and discharging of these double layers will affect the potential distribution within the cell and modify the recombination of electrons and holes. To get an idea of the timescales we are talking about, we can consider how long it takes for ionic vacancies to diffuse, say 100 nm in the case of our PSCs. Mobility values calculated for iodide ion vacancies in MAPbI$_3$ correspond to a diffusion coefficient of ca. 10^{-12} cm^2 s^{-1}. The time to diffuse a distance l is given by $t = l^2/D = 100$ s. Clearly, this is the right order of magnitude to explain the slow relaxation we are seeing in the PEIS. The low-frequency PEIS arc could then be explained if we assume that the slow relaxation of ionic distributions leads to a decrease in recombination.

This chapter ends with a note of warning about the use of equivalent circuits to model the impedance of systems that we do not understand. There is a temptation to add impedance elements to 'describe' various postulated physical processes. The addition of more elements runs into the problem of circuit degeneracy as we have seen earlier in this book. At the same time, the use of more independent variables has no merit unless there is a meaningful correlation with other diagnostic measurements. An illustration of the extent to which this approach has been taken is provided by Figure 7.23, which is a nine-element equivalent circuit proposed for fitting the PEIS response of PSCs.

An alternative approach that has been taken is to set up a detailed physical model of electron/hole–ion interaction and to calculate impedance responses numerically (see, for example, Clarke *et al.*, 2023). Of course, this

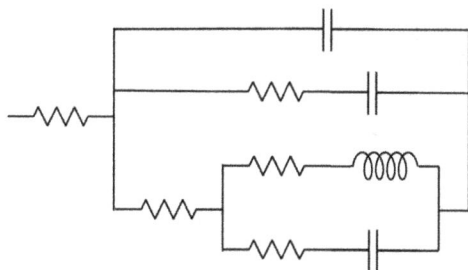

Figure 7.23. A nine-element equivalent circuit that has been proposed for fitting the PEIS response of perovskite solar cells.

'explains' the impedance response, but extraction of meaningful information from measured PEIS responses is more difficult once we abandon the equivalent circuit approach.

Appendix 7.1   How to Interpret the Resistors in the Bisquert#3 Circuit

A very clear account of the ideas summarized in Appendices 7.1 and 7.2 can be found in Peter Würfel's excellent text book *Physics of Solar Cells* (Würfel, 2005).

Metals have a large and uniform concentration of free electrons. To make a current flow in a metallic conductor, we need to change the *potential energy* of electrons as a function of position by applying a voltage. The excess potential energy of electrons due to the electric field causes them to move towards the lower potential energy of the positive contact. The electrons in a metal behave as an electron gas, which means that they are in rapid thermal motion. The presence of the electrical field adds a small velocity component to the movement of the electrons which causes them to *drift* towards the positive end of the sample while maintaining their rapid thermal motion. The *conductivity*, σ, of a material depends on n, the number of electrons per unit volume, and their *mobility* u_n – equation (A7.1).

$$\sigma = nu_n q \qquad (A7.1)$$

where q is the elementary charge. The mobility is the ratio of the velocity to the electric field, so it has units $cm\,s^{-1}/V\,cm^{-1} = cm^2\,V^{-1}\,s^{-1}$. You should be able to work out that the units of conductivity are $A\,V^{-1}\,cm^{-1} = S\,cm^{-1}$.

The current density flowing in a metallic conductor of length l depends on the number of electrons per unit volume moving with a drift velocity given by the product $u_n E = u_n V/l$.

$$j = qnE = qnu_n \frac{V}{l} \tag{A7.2}$$

Equation (A7.2) shows that the current density depends linearly on voltage because the drift velocity depends linearly on the electric field. This is the basis for Ohm's law.

It follows that the resistance, R, of some piece of material, a wire, for example, depends on its length l, its cross-sectional area A and the conductivity, σ, of the material – equation (A7.3).

$$R = \frac{l}{\sigma A} \tag{A7.3}$$

Low value resistors can be fabricated using wire wound into an insulating core. These often have unacceptably high inductance values. Higher resistance values are made using thin films of carbon, metal or metallic oxides to reduce the cross-sectional area A.

In summary, the concept of metallic resistance involves a high concentration of electrons *drifting* slowly in an electric field. So, let us look at some of the 'resistances' that we have met so far in this book. The Faradaic resistance was obtained by linearizing the Butler–Volmer equation, so using a resistor to model the current–voltage relationship is only valid for small voltage changes. Obviously, this resistor has nothing to do with conductivity. We also came across the concept of the *Warburg resistance* as the low-frequency intercept of the finite Warburg impedance. This resistance is related to the *diffusion* of electroactive species rather than to the movement of ions in an electric field. This is why we used Fick's laws to derive the impedance. The same applies to the resistors R_{tr} in the Bisquert#3 circuit. They describe *diffusion* of electrons in the mesoporous TiO_2. What we need is a way of universally describing the movement of charged or neutral species that includes both drift and diffusion. This can be done by considering the *free energy* of particles.

The *Gibbs free energy* G is widely used in thermodynamics, and it is defined as

$$G = U + PV - TS \tag{A7.4}$$

Here, U is the internal energy of the system, P and V are the pressure and volume, T and S are the temperature and entropy. G is an extensive property, i.e., it depends on the amount of material present. It is therefore convenient to define an intensive partial differential quantity called the *chemical potential*, μ_i, as

$$\mu_i = \left(\frac{\partial G}{\partial n_i}\right)_{T,P,n_j} = \mu_i^0 + k_B T \ln \left(\frac{n_i}{N_i^0}\right) \qquad (A7.5)$$

which tells us how G changes with the number of species i at constant T and P and a constant number of all other species, j (in chemistry texts n_i is defined in moles rather than numbers, in which case $k_B T$ is replaced by RT). Here, μ_i^0 is the *standard chemical potential* of species i and N_i^0 corresponds to the *standard state*.

Since we are often dealing with charged particles (electrons, ions), we need to recognize that in this case the Gibbs energy also depends on the *electrical potential*, ϕ, of the phase in which the particles are present. This potential is referred to as the inner of Galvani potential of the phase, and the electrical field $E = -d\phi/dx$. If we have a species with a charge z_i, we define the *electrochemical potential*, $\bar{\mu}_i$, as

$$\bar{\mu}_i = \mu_i + z_i q\phi = \mu_i^0 + k_B T \ln \left(\frac{n_i}{N_i^0}\right) + z_i q\phi \qquad (A7.6)$$

The logarithmic term is related to the *entropy change* at constant temperature in going from the standard state N_i^0 to the state where the number of species per unit volume is n_i. This is the same as the entropy change for the expansion of a gas at constant temperature. The additional term $z_i q\phi$ represents the work done in transporting a charge from infinity to the interior of the phase. For electrons, $z_i = -1$. The *electrochemical potential of electrons*, $\bar{\mu}_n$, is an important quantity in solid-state physics, and it is equivalent to the *Fermi energy*, E_F.

In terms of transporting species, what matters is the *gradient of free energy*. A gradient of potential energy corresponds to a *driving force*, and we can apply the same concept here. Noting that the gradient of electrochemical potential has units J cm^{-1}, and remembering that J $\equiv$ CV, allows us to see that $\frac{1}{q}\frac{d\bar{\mu}_i}{dx}$ has units V cm^{-1}, the same as those of the electrical field that we saw in equation (A7.2). We can obtain the gradient of electrochemical potential from equation (A7.6) as

$$\frac{d\bar{\mu}_i}{dx} = \frac{k_B T}{n_i}\frac{dn_i}{dx} + z_i q\frac{d\phi}{dx} \qquad (A7.7)$$

The flux of species i is therefore given by analogy with equation (A7.2).

$$J_i = -n_i u_i \frac{1}{q}\frac{d\bar{\mu}_i}{dx} = -\frac{u_i k_B T}{q}\frac{dn_i}{dx} - z_i n_i u_i \frac{d\phi}{dx} \qquad (A7.8)$$

This is nice. We have a term that depends on the gradient $\frac{dn_i}{dx}$ and a term that depends on $\frac{d\phi}{dx}$. The first term must therefore be Fick's first law in disguise, and indeed it is because if you check the units, you will see that $n_i u_i / q = D_i$, the diffusion coefficient of spices i. You should recognize the $d\phi/dx$ terms as the negative of the electrical field, and so the second term is obviously related to the movement of charged species in an electrical field – it is the *drift* process that we saw earlier. We see that using the free energy we have an expression for the flux (or current density if we multiply J_i by $z_i q$) that contains contributions from both diffusion and drift. It is important to understand that this *separation into diffusion and drift is entirely notional* – there is no experimental way that we can separate the drift and diffusion components in the flux.

Now we can return to our questions about the meaning of the transport resistors in the Bisquert#3 circuit. In the mesoporous TiO$_2$, the electric field is negligible, so the drift term in equation (A7.8) disappears. Multiplying by q and setting $z_i = -1$ to obtain the current density, we find

$$j = -q n_i u_i \left(\frac{1}{q} \frac{d\bar{\mu}_i}{dx} \right) \tag{A7.9}$$

Check the units of the term in brackets – you should find that they are volts cm^{-1}. The units of $q n_i u_i$ are A V^{-1} = S cm^{-1}. We can rewrite equation (A7.9) in the form of Ohm's law for a small element of length δx

$$-\left(\frac{1}{q} \frac{d\bar{\mu}_i}{dx} \right) \delta x = j \left(\frac{1}{q n_i u_i} \right) \delta x \tag{A7.10}$$

The left-hand side now has unit of volts and the term in brackets on the right-hand side has units of – yes, you guessed it – Ω cm^2. Multiplying by the area of the TiO$_2$ layer, we have discovered that the electron transport resistance in the Bisquert#3 circuit is

$$R_{\text{tr}} = \left(\frac{A}{q n_i u_i} \right) \delta x \tag{A7.11}$$

Appendix 7.2   Fermi Levels and Quasi-Fermi Levels (QFLs)

The probability of electron occupation for an energy level E is given by the *Fermi–Dirac* equation.

$$f_{\text{FD}} = \frac{1}{1 + \exp\left(\frac{E - E_F}{k_B T} \right)} \tag{A7.12}$$

For $(E - E_F)/k_B T \gg 1$, the Fermi–Dirac equation reduces to the Boltzmann equation.

$$f = \exp\left(-\frac{E - E_F}{k_B T}\right) \tag{A7.13}$$

In this limit, the equilibrium (dark) concentrations of electrons and holes are given by

$$n = f_{\mathrm{FD}} N_C = N_C e^{-\left(\frac{E_C - E_F}{k_B T}\right)} \tag{A7.14a}$$

$$p = (1 - f_{\mathrm{FD}}) N_V = N_V e^{\left(\frac{E_F - E_V}{k_B T}\right)} \tag{A7.14b}$$

Here, N_C, N_V are the *effective density of states* in the conduction and valence bands, which are located at energies E_C and E_V, respectively. The Fermi energy E_F is a free energy that is equivalent to the *electrochemical potential* of electrons (see Appendix 7.1).

At thermal equilibrium in the dark, the Fermi energy is the same everywhere in a solar cell, including in the contacts. It follows that the gradient of electron free energy is zero everywhere so that there is no driving force for the movement of electrons. Under illumination at open circuit, electrons are promoted from the valence band to the conduction band and subsequently return to the valence band by electron–hole *recombination*. This process can be radiative, with emission of photons, or non-radiative, with the emission of phonons (lattice vibrations). Under steady-state conditions, the rates of electron–hole pair creation and recombination are equal and the steady-state electron and hole concentrations in the absorber are $n + \Delta n$ and $p + \Delta p$, where $\Delta n = \Delta p$. The excess electrons and holes are in thermal equilibrium with the lattice, allowing us to use the steady-state electron and hole concentrations to define separate *quasi-Fermi* levels for electrons and holes.

$$n + \Delta n = N_C e^{-\left(\frac{E_C - {}_n E_F}{k_B T}\right)} \tag{A7.15a}$$

$$p + \Delta p = N_V e^{-\left(\frac{{}_p E_F - E_V}{k_B T}\right)} \tag{A7.15b}$$

In an ideal solar cell, the splitting of the QFLs is reflected in the *open-circuit voltage*, V_{oc}.

$$nE_F - pE_F = qV_{oc} \tag{A7.16}$$

In addition to the absorber, a solar cell needs *selective contacts* that allow one carrier type to pass but block the other. An n-doped semiconductor contact has a high conductivity for electrons, while a p-doped semiconductor has high conductivity for holes, so we can make a device in which a low doped (or undoped) absorber is sandwiched between a p-type contact and an n-type contact. If the absorber is undoped (insulating), we have a p-i-n solar cell. Figure A7.1 illustrates the band and Fermi energy alignments for such a p-i-n solar cell (actually an 'n-i-p' solar cell if we read it from left to right). In this case, both contact materials have higher band gaps than the absorber, so this is an example of a *heterojunction cell*. If the contact and absorber are the same material (e.g., silicon), the device is a homojunction solar cell. If the absorber layer is omitted, we have a *p-n junction solar cell* in which the distinction between absorber and selective contact is less clear-cut.

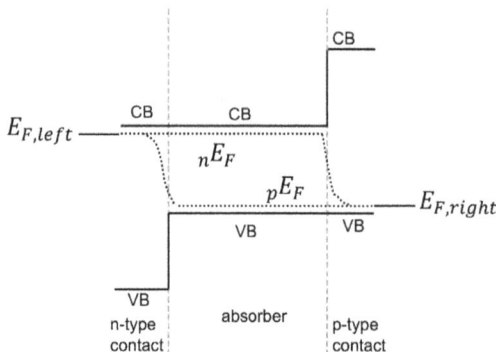

Figure A7.1. Energy band diagram of an (idealized) illuminated p-i-n solar cell showing the selective contacts and the splitting of the quasi-Fermi levels (QFLs) and the resulting generation of an *open-circuit voltage*, which corresponds to the difference between the Fermi levels of the contacts on the two sides. The left-hand contact is a wide band gap n-type semiconductor that allows easy movement of conduction band electrons to the left, but blocks hole movement. Note the step in the valence band energy that makes it hard for electrons to move from the valence band of the contact to the valence band of the absorber (this corresponds to the transfer of a hole from the absorber to the contact). The right-hand contact is a wide band gap p-type semiconductor that allows holes to move to the right but blocks electrons.

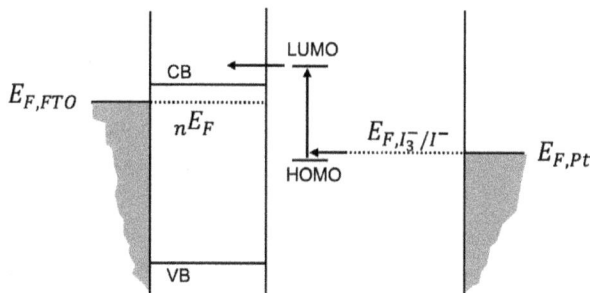

Figure A7.2. Energy level alignments in the DSSC. Note that we have a *molecular absorber*, so that the relevant energy levels are the occupied (HOMO) and unoccupied (LUMO) levels of the dye rather than the valence and conduction band energies of a conventional absorber. The TiO$_2$ acts as an electron-selective contact and the iodide electrolyte, as the hole-selective contact.

Having looked at an idealized solar cell, we can now ask whether the two types of cells that we are concerned with in this chapter fit the picture illustrated in Figure A7.1. The simplest cell to consider is the planar PSC in which a thin layer of metal halide perovskite is sandwiched between an n-type TiO$_2$ layer and 'p-type' organic conductor such as spiro-MeOTAD (2,2',7,7'-Tetrakis[N,N-di(4-methoxyphenyl)amino]-9,9'-spirobifluorene).

This configuration does correspond reasonably well to Figure A7.1, although the spiro-MeOTAD is not a conventional p-type semiconductor. Its oxidized form is more like the highly conducting emeraldine salt of PANI that is discussed in Chapter 8. At first sight, the DSSC does not seem to fit the description in Figure A7.1. However, once we recognize that the absorber here is the dye, it becomes clear that the mesoporous TiO$_2$ is the electron-selective contact, whereas the iodide electrolyte is the hole-selective contact. Figure A7.2 illustrates the alignment of energy levels and the origin of the open-circuit voltage in this case.

Appendix 7.3 The Chemical Capacitance

The second element in the Bisquert#3 circuit that we need to discuss is the *chemical capacitance*, C_{chem}. This concept relates to the way in which the electronic charge in a system varies as the electrochemical potential or Fermi energy changes. The chemical capacitance is given by

$$C_{\text{chem}} = \frac{dQ}{\frac{1}{q}d\bar{\mu}_n} = \frac{qdn}{\frac{1}{q}d\bar{\mu}_n}Adx = q^2\frac{dn}{d\bar{\mu}_n}Adx = q^2\frac{dn}{dE_F}Adx \qquad (A7.17)$$

Note that the $1/q$ term converts the electrochemical potential of Fermi energy into a voltage (remember $J \equiv CV$) so that the units of the chemical capacitance as defined by equation (A7.12) are $CV^{-1} = F$. Further, note that the chemical capacitance is proportional to the volume of the material – in equation (A7.12) this is Adx, where A is the area and dx is the increment in thickness.

We have two types of electronic charges to consider when discussing the chemical capacitance. The first is the charge associated with electrons and holes that are free to move in the conduction and valence bands, respectively. The second is the charge associated with electrons or holes trapped in energy states that lie in the forbidden energy gap of the semiconductor. We will begin by assuming we have an ideal semiconductor with no trap states. If the Fermi energy is sufficiently far away that we can use the Boltzmann limit of the Fermi–Dirac equation – see equation (A7.13) – it is easy to show from equation (A7.14) that the chemical capacitance per unit volume associated with conduction band electrons is given by

$$C_{\text{chem,CB}} = \frac{q^2 n}{k_B T} = \frac{q^2 N_C}{k_B T} \cdot e^{-\frac{(E_c - E_F)}{k_B T}} \qquad (A7.18)$$

To obtain an idea of how large the chemical capacitance might be in an illuminated solar cell without traps, we will suppose that we have an undoped material with a band gap of 1.5 eV and an effective density of states of $N_C = 10^{19}$ cm^{-3} – both values that are typical for perovskite solar cells. If the Fermi level splitting $= 1$ eV, corresponding to an open-circuit voltage of 1 V, is symmetrical about the centre of the band gap, then $(E_c - E_F)$ in equation (A7.18) would be half of the difference between

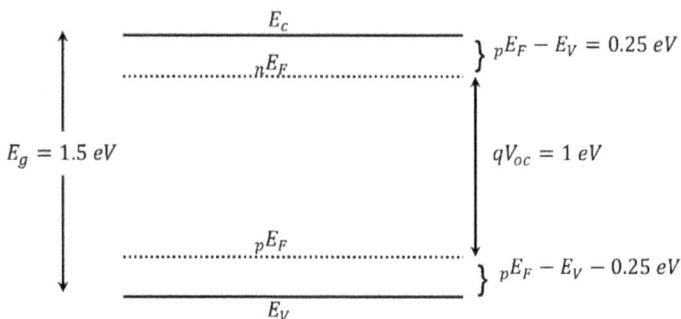

Figure A7.3. Idealized Fermi level splitting in a solar cell allowing calculation of the conduction band chemical capacitance. Here, the band gap is 1.5 eV and the Fermi level splitting under illumination is 1 V, both values that are typical for perovskite solar cells.

the band gap and the Fermi level splitting as illustrated in Figure A7.3, i.e., 0.25 eV. Substituting these values, we find that the electron concentration in the conduction band is 6×10^{13} cm^{-3}, and the corresponding chemical capacitance per unit volume is 3.7×10^{-4} F cm^{-3}. This means that if we have a planar perovskite solar cell that is 500 nm thick, the capacitance would be 3.7×10^{-4} F cm$^{-3} \times 5 \times 10^{-5}$ cm $= 18.5$ nF cm^{-2}. The *geometric* capacitance of the same PSC calculated using a dielectric constant of 24 is 42 nF cm^{-2}. So, C_{chem} and C_{geo} are quite similar in this case. We expect the chemical capacitance to dominate at higher intensities (higher open-circuit voltages) or for thicker perovskite layers.

We turn now to the DSSC. Under illumination at open circuit, the electron concentration in the mesoporous TiO$_2$ layer reaches very high values. Most of these electrons are trapped at states located in the forbidden energy gap of the TiO$_2$. The trapped electrons can be released back to the conduction band by absorption of thermal energy, and under steady-state conditions, the rates of trapping and detrapping are equal. The ability of the mesoporous TiO$_2$ to store charge allows us to define a chemical capacitance associated with trapped electrons. The trap states appear to be distributed exponentially in energy as shown in Figure A7.4. The distribution of traps in the band gap of the mesoporous TiO$_2$ is described by the density of

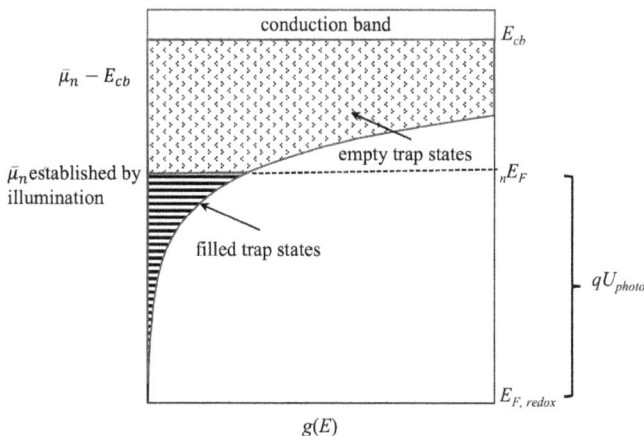

Figure A7.4. Exponential density of states function for electron trapping states in mesoporous TiO$_2$ corresponding to equation (A7.19). The left-hand scale is in terms of the electrochemical potential. The equivalent right-hand scale is in terms of the Fermi energy. Illumination establishes a photovoltage that is given by the difference between the redox Fermi energy, $E_{F,\text{redox}}$, and the quasi-Fermi energy, $_nE_F$, of electrons in the TiO$_2$. Trap states below $_nE_F$ are filled with electrons, those above $_nE_F$ are empty. The chemical capacitance of the DSSC is mainly due to trapped electrons.

states function

$$g(E) = \frac{\alpha N_{\text{trap}}}{k_B T} e^{\frac{\alpha(E - E_{\text{cb}})}{k_B T}} \tag{A7.19}$$

The distributed chemical capacitance elements associated with the trapped electrons are given by

$$C_{\text{chem}}^{\text{traps}} = q^2 \frac{\alpha N_{\text{trap}}}{k_B T} e^{\frac{\alpha(E_F - E_{\text{cb}})}{k_B T}} A dx \tag{A7.20}$$

References

Clarke, W., Bennett, L. J., Grudeva, Y., Foster, J. M., Richardson, G. & Courtier, N. E. 2023. IonMonger 2.0: Software for free, fast and versatile simulation of current, voltage and impedance response of planar perovskite solar cells. *Journal of Computational Electronics*, 22, 354–382.

Guerrero, A., Bisquert, J. & Garcia-Belmonte, G. 2021. Impedance spectroscopy of metal halide perovskite solar cells from the perspective of equivalent circuits. *Chemical Reviews*, 121, 14430–14484.

Juarez-Perez, E. J., Sanchez, R. S., Badia, L., Garcia-Belmonte, G., Kang, Y. S., Mora-Sero, I. & Bisquert, J. 2014. Photoinduced giant dielectric constant in lead halide perovskite solar cells. *Journal of Physical Chemistry Letters*, 5, 2390–2394.

Kojima, A., Teshima, K., Shirai, Y. & Miyasaka, T. 2009. Organometal halide perovskites as visible-light sensitizers for photovoltaic cells. *Journal of the American Chemical Society*, 131, 6050–6051.

Oregan, B. & Gratzel M. 1991. A low-cost, high-efficiency solar-cell based on dye-sensitized colloidal TiO$_2$ films. *Nature*, 353, 737–740.

Pockett, A. 2017. *Characterization of Perovskite Solar Cells*. PhD Thesis, University of Bath.

Pockett, A., Eperon, G. E., Peltola, T., Snaith, H. J., Walker, A., Peter, L. M. & Cameron, P. J. 2015. Characterization of planar lead halide perovskite solar cells by impedance spectroscopy, open-circuit photovoltage decay, and intensity-modulated photovoltage/photocurrent spectroscopy. *The Journal of Physical Chemistry C*, 119, 3456–3465.

Wang, H. X. & Peter, L. A. 2009. A comparison of different methods to determine the electron diffusion length in dye-sensitized solar cells. *Journal of Physical Chemistry C*, 113, 18125–18133.

Würfel, P. 2005. *Physics of Solar Cells: From Principles to New Concepts*. Weinheim: Wiley.

Chapter 8

Electrochromic Systems: Potential-Modulated Absorbance Spectroscopy of Polyaniline, and Light-Modulated Absorbance of Haematite

8.1 Introduction

Potential-modulated absorbance and reflectance spectroscopy (PMAS/PMRS) techniques have been developed to detect species formed by electrochemical redox reactions. Their high sensitivity arises from the use of lock-in detection of the modulated optical signal: changes as small as one part per million in absorbance or reflectance can be measured with relative ease. In this chapter, we introduce the concept of *complex capacitance* and show how it is related to the PMAS response of an electrochromic layer of the conducting polymer, polyaniline, deposited in a conducting glass electrode by the electropolymerization of aniline. The PMAS technique has also been used to study haematite (α-Fe_2O_3) electrodes used for light-driven water splitting (photoelectrolysis). We compare the PMAS results on haematite with results obtained using the closely related technique, *light-modulated* absorbance spectroscopy (LMAS), which probes the creation and decay of higher valent Fe surface intermediates formed during the photoelectrochemical oxygen evolution reaction (POER). In LMAS, the illumination intensity of the single wavelength UV exciting light is modulated rather than the electrode potential, which is held constant.

The modulated absorbance is measured at longer wavelengths using a blocking filter and monochromator to remove stray UV excitation light.

8.2 Polyaniline Electrochromic Window

The 2000 Nobel Prize in Chemistry was awarded jointly to Alan J. Heeger, Alan G. MacDiarmid and Hideki Shirakawa for the discovery of *conductive polymers*. These novel materials combine the properties of plastics and metals in a unique way. Heeger and MacDiarmid pioneered studies of polyaniline (PANI), which is a conducting polymer that can be synthesized by chemical or electrochemical oxidation of aniline (Shirakawa discovered polyacetylene). PANI films can be grown by *electropolymerization* on conducting glass (fluorine-doped tin oxide: FTO coated) electrodes by potential cycling in an acidic solution containing the aniline monomer. In this work, transparent FTO-coated glass (TEC-15) was used as a working electrode.

Depending on the electrode potential, electropolymerized PANI layers can be switched between a reduced transparent *leuco* form and an oxidized green *emeraldine salt* form by potential cycling in acidic solution. The PANI layer on FTO-coated glass can therefore act as an *electrochromic window*. Figure 8.1 shows the cyclic voltammogram of the electropolymerized film

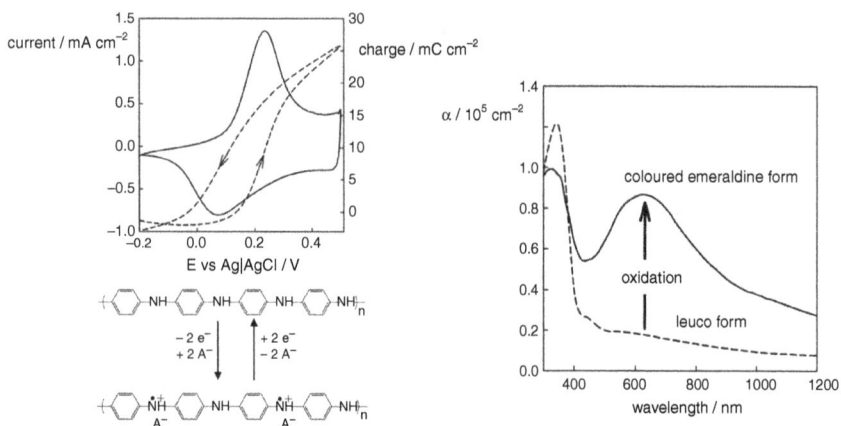

Figure 8.1. Cyclic voltammogram showing cycling of PANI film deposited by electropolymerization on conducting glass in 1.0 M H_2SO_4 (sweep rate 10 mV s^{-1}). Oxidation of the *leuco* form generates the radical ion salt or *emeraldine* form. The leuco form is transparent and insulating, whereas the emeraldine salt is green and has very high conductivity. The absorption spectra of the two forms are shown on the right.

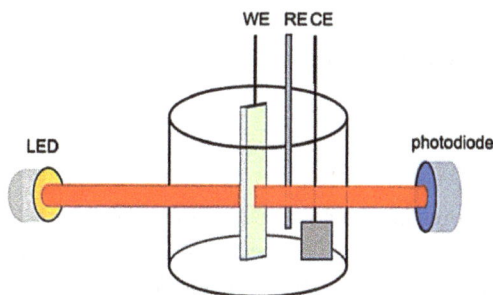

Figure 8.2. Simple experimental setup for frequency-resolved PMAS on electrochromic PANI films. The modulated transmittance of the PANI film is measured using a collimated red LED (633 nm) and a silicon photodiode. The photodiode is connected via a fast current/voltage converter to one channel of the frequency response analyzer (not shown), while the other FRA channel measures the modulating voltage.

used in the present study as well as the structure of the two forms of PANI and the electrochromic switching reaction.

8.3 A Simple Experimental Setup for Frequency-Resolved PMAS

PMAS is normally carried out using a lamp and monochromator so that the spectral response can be measured at a constant modulation frequency. Our optical setup for frequency-resolved measurements is much simpler (Kalaji and Peter, 1991). We just use a red LED (633 nm), since this wavelength is close to the maximum absorbance of the emeraldine form. The light transmitted through the PANI film is detected by a silicon photodiode placed on the other side of the electrochemical cell. The photodiode current is measured using a fast current amplifier, and the output is sent to one input of the frequency response analyzer, which measures the ratio of the AC component of the photodiode signal to the modulating voltage while varying the frequency. The optical setup is illustrated in Figure 8.2.

8.4 The Modulated Transmittance

The transmittance T of light through a solid layer of thickness d is described by the Beer–Lambert law, as follows:

$$T = \frac{I_t}{I_0} = e^{-\alpha d} \tag{8.1}$$

Here, I_0 and I_t are the incident and transmitted light intensities, and a (cm^{-1}) is the absorption coefficient. Since we have two different absorbing species (reduced: leuco and oxidized: emeraldine) in our electrochromic PANI layer, Equation (8.1) needs to be rewritten as

$$T = \frac{I_t}{I_0} = e^{-(f_O a_O + f_R a_R)} \tag{8.2}$$

Here, f_O and f_R represent the volume fractions of oxidized (emeraldine) and reduced (leucoemeraldine) PANI species, and a_O, a_R are the corresponding absorption coefficients at a fixed wavelength (633 nm in our case). Note that the sum of f_O and f_R must be 1, which will help us in our derivation of a final expression for the *modulated transmittance*, which we shall call ΔT.

Theory Note 8.1. The Modulated Transmittance

In the PMAS experiment, we fix a DC electrode potential and modulate the potential sinusoidally by a small amount δV. This potential modulation changes the fractions of oxidized and reduced PANI species by an amount $\delta f_O = -\delta f_R$. It follows from equation (8.2) that the corresponding change in transmittance ΔT is given by

$$T + \Delta T = e^{-[(f_O + \delta f_O)a_O + (f_R - \delta f_O)a_R]d} = e^{-[(f_O a_O + f_R a_R) + \delta f_O(a_O - a_R)]d} \tag{8.3}$$

Remembering that $e^{ab} = e^a \cdot e^b$, we can split up the exponent to give

$$T + \Delta T = e^{-(f_O a_O + f_R a_R)d} \cdot e^{-[\delta f_O(a_O - a_R)d]} = T e^{-[\delta f_O(a_O - a_R)d]} \tag{8.4}$$

Now comes the linearization that we have seen several times in previous chapters where we replace e^x by $1 + x$ for small x. For a small voltage modulation amplitude, we can replace the $e^{-[\delta f_O(a_O - a_R)d]}$ term in equation (8.4) by $1 - \delta f_O(a_O - a_R)d$ to give

$$T + \Delta T = T[1 - \delta f_O(a_O - a_R)d] = T - T[\delta f_O(a_O - a_R)d] \tag{8.5}$$

Thus, we see that the *normalized change in transmission* is

$$\frac{\Delta T}{T} = \delta f_O(a_O - a_R)d \tag{8.6}$$

Now let us consider how the change in the emeraldine fraction is related to the charge passed during the potential modulation. If the charge required to completely oxidize the film from the leuco form to emeraldine is Q_0, we can replace δf_O in equation (8.6) by $\delta Q/Q_0$ to give

$$\frac{\Delta T}{T} = \frac{(a_O - a_R)d}{Q_0} \delta Q \tag{8.7}$$

The purpose of this derivation was to show that *the PMAS response is linearly related to the AC charge passed during modulation.* We now need to work out how this AC charge δQ is related to the impedance of the film.

8.5 The Impedance, Admittance, and Complex Capacitance of PANI Films

PANI films can store charge reversibly as shown in the charge plot in Figure 8.1. In fact, they behave like *supercapacitors.* In this example, the film stores a charge of 25 mC cm^{-2} for a potential change of around half a volt. This corresponds to a very high capacitance of around 50 mF cm^{-2} (remember $Q = CV$ for an ideal capacitor). In fact, Figure 8.1 shows that the charge does not vary linearly with voltage, but for small changes in potential we can linearize the charge/voltage plot and treat the film as a capacitor. This capacitor is charged and discharged through a series resistance that is mainly due to the resistance of the conducting glass (electron transfer is sufficiently fast that the speed of charging is entirely determined by the RC time constant). The magnitude of the charge modulation δQ (and hence the magnitude of the modulated transmittance) is therefore frequency-dependent because the total modulation voltage is split into a part across the series resistance and a part across the capacitance. At very low frequencies, the impedance is dominated by the capacitor so that the voltage appears almost entirely across the PANI electrode, but at high frequencies the opposite is true and nearly all the modulation occurs across the series resistance. The key factor here is the time constant RC of the circuit. You will remember that for $\omega = 1/RC$, half of the AC voltage will appear across R and the other half across C in the case of a series RC circuit.

We will now investigate the consequences of the series RC circuit for the electrochromic modulation measured by PMAS. We begin with a Theory Note to discover something called the *complex capacitance* (Theory Note 8.2). This quantity can be displayed in ZView$^{®}$ by choosing the axes as E', E'' (more correctly ε' and ε'' – ZView$^{®}$ does not use Greek symbols) which are referred to as 'complex dielectric'. This is because the capacitance of a dielectric capacitor is related to dielectric constant (also called relative permittivity) by

$$C = \frac{A\varepsilon\varepsilon_0}{d} \tag{8.8}$$

Here, ε is the dielectric constant or relative permittivity of the dielectric between the capacitor plates, ε_0 is the permittivity of free space (8.8542×10^{-14} F cm^{-1}) and d is the separation between the capacitor plates, which have area A. This means that ε and C are linearly related.

Theory Note 8.2. The Complex Capacitance

Our equivalent circuit for the PANI electrode is a series RC connection as shown in Figure 8.3. It has an impedance given by

$$Z = R - \frac{j}{\omega C} \qquad (8.9)$$

Since we are interested in the AC component of the electrode charge, it is more convenient to consider the AC current, which depends linearly on the admittance. You should be able to show that the admittance Y of the series RC circuit is given by

$$Y = \frac{R\omega^2 C^2}{1 + \omega^2 R^2 C^2} + j\frac{\omega C}{1 + \omega^2 R^2 C^2} \qquad (8.10)$$

The AC current is given by

$$\delta i = Y \delta V = Y |\delta V| e^{i\omega t} \qquad (8.11)$$

We are interested in the AC charge modulation, so we need to *integrate the current* with respect to time.

Integrating the AC current gives us the AC charge.

$$\delta Q = \int \delta i \cdot dt = \int Y \delta V_0 e^{j\omega t} \cdot dt = \frac{Y}{j\omega}\delta V_0 e^{j\omega t} = -j\frac{Y}{\omega}\delta V \qquad (8.12)$$

This equation looks a bit like the well-known capacitor equation $Q = CV$, so we can write it as

$$\delta Q = \hat{C}\delta V \qquad (8.13)$$

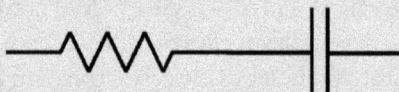

Figure 8.3. Series RC circuit used to model the PMAS response of the PANI electrochromic window. The series resistance is mainly due to the FTO. The (pseudo)capacitance of the PANI film is in the mF cm^{-2} range, giving a long time constant for electrochromic switching.

Theory Note 8.2. (*Continued*)

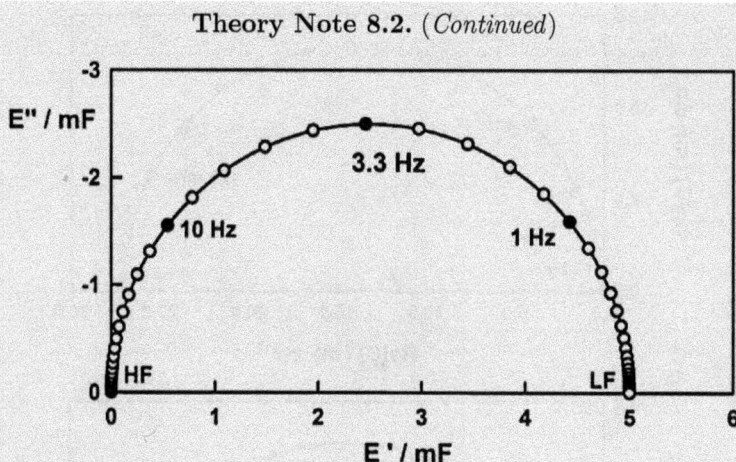

Figure 8.4. 'Complex dielectric' plot of $\hat{C}$ for a series RC circuit. $R = 10\ \omega$, $C = 5$ mF. Note $\omega_{max} = \frac{1}{RC}$.

where

$$\hat{C} = -j\frac{Y}{\omega} \qquad (8.14)$$

is the *complex capacitance*.

Now we can go back to equation (8.10) to obtain the real and imaginary components of the complex capacitance. You should be able to show that

$$\hat{C} = \frac{C}{1 + \omega^2 R^2 C^2} - j\frac{R\omega C^2}{1 + \omega^2 R^2 C^2} \qquad (8.15)$$

In the low-frequency limit, $\hat{C} = C$, whereas $\hat{C}$ tends to zero in the high-frequency limit. You can see from equation (8.15) that the real and imaginary components of $\hat{C}$ are equal when $\omega = 1/RC$. The complex capacitance therefore traces a semicircle in the complex plane as shown in Figure 8.4. Note that $\omega_{max} = 1/RC$.

The complex capacitance concept discussed above applies to the PANI electrode. The only difference is that the charging of the PANI *pseudo-capacitance* takes place by oxidation/reduction reactions rather than by charging a dielectric. Now we can return to equation (8.7), which related the PMAS response to the modulated charge δQ. We replace δQ by $\hat{C}\delta V$ (the capacitor equation again) to obtain

$$\frac{\Delta T}{T} = \frac{\hat{C}(\alpha_O - \alpha_R)d}{Q_0}\delta V \qquad (8.16)$$

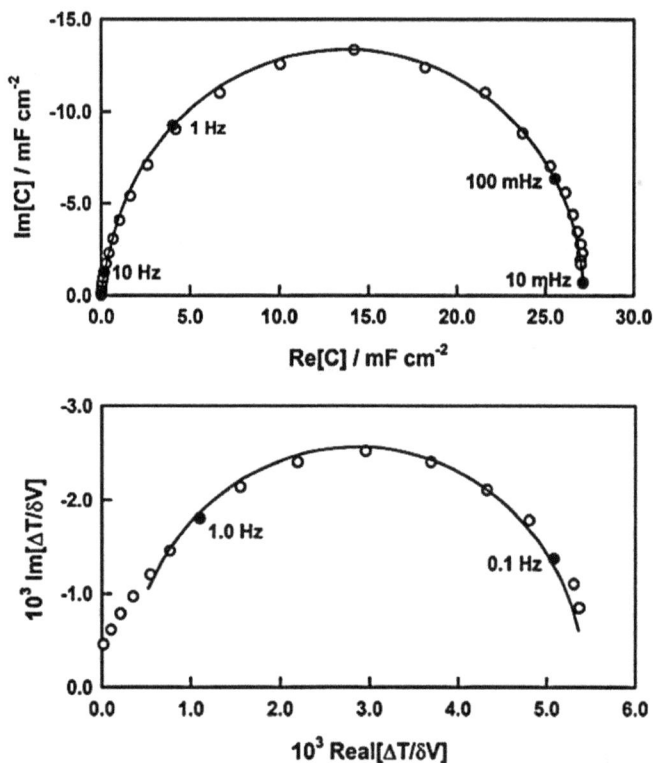

Figure 8.5. Comparison of Nyquist plots of the complex capacitance measured by EIS with the frequency-resolved PMAS response of PANI film. Note the very close correspondence. Both measurements were made at a potential of 0.4 V vs. Ag/AgCl.

which we can rearrange to give

$$\frac{\Delta T}{T \delta V} = \frac{\hat{C}(\alpha_O - \alpha_R)d}{Q_0} \tag{8.17}$$

The ratio $\Delta T/\delta V$ is what we measure with our frequency response analyzer (remember that δV is the AC modulation voltage and ΔT is the AC PMAS signal). T is the DC transmittance, which we measure as the DC component of the photodiode output. Since the terms T, α_O, α_R, Q_0 and d in equation (8.17) are all constants, this result shows that the frequency dependence of the PMAS signal should exactly mirror the frequency dependence of the complex capacitance $\hat{C}$. Figure 8.5 confirms this conclusion. In practice, we normally plot the normalized transmittance change $\Delta T/T$ rather than $\Delta T/T\delta V$ since this is what appears in the PMAS spectra.

The experimental results confirm that the time constants of the electrical and optical responses are identical. The PMAS response at high frequencies deviates from the semicircle because the RC circuit is an oversimplification. In fact, the resistance of the conducting glass electrode is distributed across the electrode area under the polyaniline so that a finite transmission line circuit is more appropriate (see Kalaji and Peter, 1991).

The same approach can be used for frequency-resolved reflectance measurements for non-transparent samples. In this case, the method is referred to as potential-modulated reflectance spectroscopy (PMRS). The high sensitivity of the lock-in detection means that PMRS can be used to study sub-monolayer films on electrodes with the option of combining full spectral resolution with frequency response analysis. In the next section, we look at PMAS experiments on thin layers of haematite (α-Fe$_2$O$_3$) before describing light-modulated absorption spectroscopy (LMAS) on the same films.

8.6 PMAS of Mesoporous Haematite Films

Electrolysis offers a green route to hydrogen, provided that renewable energy is used to generate the electricity. Solar or wind-generated electrical power can be used for conventional electrolysis, but a tempting alternative is to use *photoelectrolysis*. This process involves using light-absorbing photoelectrodes to drive water splitting (i.e., hydrogen and oxygen evolution) directly. For the photoelectrochemical oxygen evolution reaction (POER), stable oxides such as haematite (α-Fe$_2$O$_3$) that absorb in the visible spectrum have been widely studied. POER is a four-electron process that involves photogenerated intermediates that are bound to the electrode surface. Identifying these intermediates *in operando* is a major challenge. Here we will look at how PMAS and LMAS have been used to identify intermediates on haematite surfaces and to link their formation and decay to the impedance response measured under illumination (photoelectrochemical impedance spectroscopy – PEIS). Since we are interested in spectra as well as the frequency response of PMAS, the experimental setup is less primitive than the one used for the study of PANI electrochromic windows. The LED is replaced by a white light source and the transmitted light is passed through a monochromator before reaching the photodiode detector. The output of the photodiode has a DC component that corresponds to the "DC" transmittance T and an AC component that corresponds to the modulated transmittance ΔT.

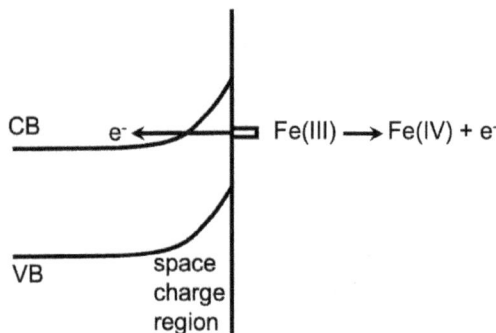

Figure 8.6. At high overpotentials in the dark, electron tunnelling from surface Fe(III) atoms to the conduction band of Fe_2O_3 creates Fe(IV) states that are intermediates in the oxygen evolution reaction.

Normalized PMAS spectra are plots of $\Delta T/T$ vs. wavelength. Since the changes in absorbance are rather small, a modulation amplitude of 100 mV was used for the PMAS experiments to obtain noise-free spectra.

The mesoporous haematite films used in this joint study between the universities of Bath and Loughborough were fabricated by a layer-by-layer process that involved repeatedly dipping FTO glass into a stable colloidal solution of hydrous oxide nanoparticles using hydrocellulose as a binder (Cummings *et al.*, 2012). The dried multilayer films were heated in air at 500°C to oxidize organic matter and to form the haematite phase. The high internal surface area of the mesoporous film increases the absorbance due to surface-bound species, making it easier to detect surface species formed in the dark (PMAS) or under illumination (LMAS).

In the dark, oxygen evolution on haematite electrodes requires a large overpotential of around 500 mV. This is because haematite is an n-type semiconductor, i.e., the majority charge carriers are electrons. Oxidation of water requires holes in the valence band, and these are not available in the dark. If the voltage becomes high enough, electron tunnelling can occur from atoms in the surface as shown in Figure 8.6. This leads to oxygen evolution via a multistep electron transfer process.

To detect surface species formed during oxygen evolution in the dark, we choose a DC potential in the region where the current begins to rise and then apply a voltage modulation and look for a modulation of the transmittance with the PMAS setup. Oxygen evolution on the haematite electrodes in 0.1 M NaOH begins at around 0.7 V vs. SCE. This corresponds to about 0.5 V vs. the reversible oxygen electrode potential (1.23 V vs. SHE)

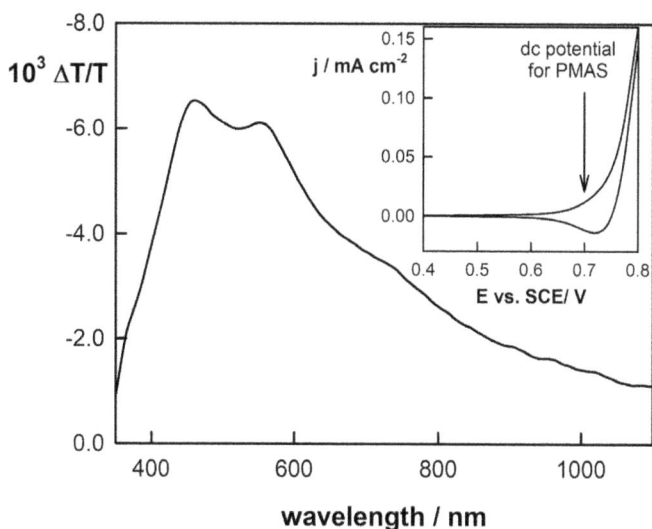

Figure 8.7. PMAS spectrum recorded at 0.7 V vs. SCE for a haematite electrode in 0.1 M NaOH in the dark. Modulation frequency 2.7 Hz. 100 mV p-p. The inset shows the cyclic voltammogram recorded at 100 mV s^{-1}. Note that the DC potential for PMAS was chosen to be in the region where oxygen evolution begins (cf. the arrow on the cyclic voltammogram).

at pH 13. The inset in Figure 8.7 shows the cyclic voltammogram of the haematite electrode in the dark, indicating the DC potential used. The PMAS spectrum in Figure 8.7 shows how the potential modulation changes the transmittance as the results of the formation of higher valent Fe species on the electrode surface. It is now known that these species are surface-bound Fe=O, which corresponds to Fe(IV). This spectrum was obtained at a modulation frequency of 2.7 Hz, since the signal attenuates at higher frequencies. This shows that the intermediate species is quite long-lived, indicating that it is involved in the slowest step in the 4-electron oxidation of water.

To find out more about the behaviour of the mesoporous haematite electrodes in the dark, EIS measurements were performed to allow correlation with the frequency-resolved PMAS response. Measurements were made in the dark at a potential close to where the largest PMAS response was observed. The mesoporous nature of the film is illustrated by the impedance of a 40-layer film shown in Figure 8.8. The Nyquist plot shows the typical transmission line behaviour at high frequencies and a low-frequency semicircle due to the parallel combination of a capacitance and

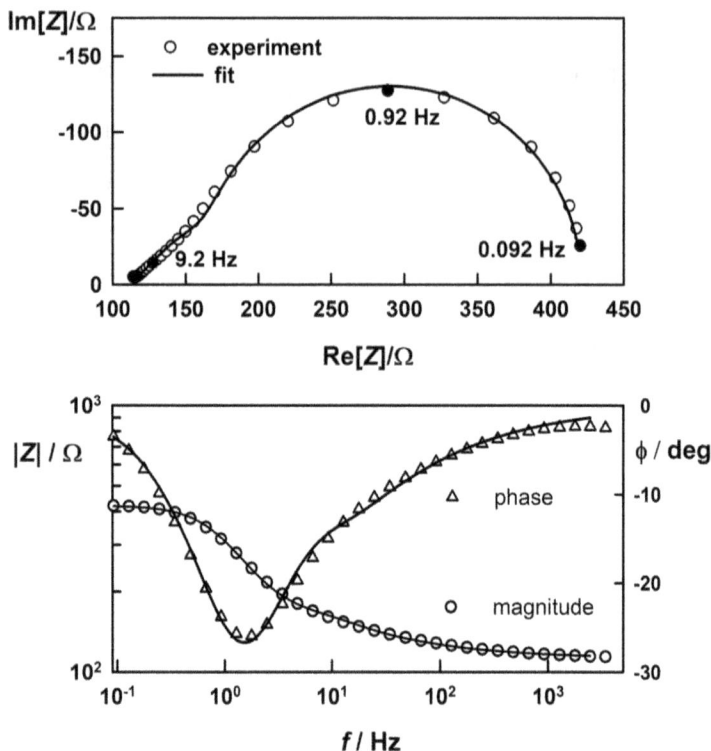

Figure 8.8. Nyquist and Bode plots for a mesoporous haematite film measured in the dark. Electrolyte 0.1 M NaOH. Potential 0.775 V vs. SCE. Electrode area 1 cm². The experimental data were fitted using the Bisquert#3 model discussed in Chapter 7.

a charge transfer resistance. The impedance data were fitted in ZView® using the Bisquert#3 model, giving $R_{tr} = 161\Omega$ cm², $R_{ct} = 257\Omega$ cm², $C_{chem} = 0.68$ mF cm⁻² (the Bisquert#3 equivalent circuit was used to model the photoelectrochemical impedance of dye-sensitized solar cells – see Chapter 7 for a full discussion). The rather high value of the transport resistance suggests that electron transport in the nanocrystalline Fe_2O_3 film occurs by hopping.

As we shall see, comparison of the impedance and PMS responses indicates that the chemical capacitance corresponds to the storage of charge in Fe(IV) surface states, i.e., to a pseudocapacitance. The charge stored in the Fe(IV) intermediate state can be 'discharged' through R_{ct}, which represents the subsequent transfer of another 3 electrons to complete the 4-electron oxidation of water. In alkaline solutions, where the OH groups on

Fe atoms are ionized, a probable mechanism for the overall oxygen evolution process in the dark is

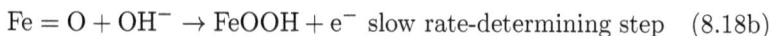

$$\text{Fe-O}^- \rightleftharpoons \text{Fe} = \text{O} + e^- \tag{8.18a}$$

$$\text{Fe} = \text{O} + \text{OH}^- \rightarrow \text{FeOOH} + e^- \quad \text{slow rate-determining step} \tag{8.18b}$$

$$\text{Fe-OOH} + 2\text{OH}^- \rightarrow \text{Fe-OH} + \text{H}_2\text{O} + 2e^- \quad \text{fast} \tag{8.18c}$$

The build-up of detectable amounts of Fe=O species indicates that the second step in the reaction sequence – equation (8.18b) – is slow and therefore probably the rate-determining step.

Equation (8.12) in Theory Note 8.2 shows that the modulated charge is proportional to the complex capacitance. For convenience, here is the expression again:

$$\delta Q = -j\frac{Y}{\omega}\delta V \tag{8.19}$$

This means we can obtain the modulated charge as a function of frequency from the admittance of the haematite electrode. If we do this using the same modulation amplitude as for the frequency-resolved PMAS, we can directly compare the frequency response of the charge and the normalized absorbance change as shown in Figure 8.9. The close correspondence between the two Nyquist plots tells us that there is a direct link between the charge passed and the change in absorbance.

8.7 Light-Modulated Absorbance Spectroscopy

We have seen that PMAS can be used to detect the long-lived intermediates formed during oxygen evolution on haematite in the dark. However, the real interest in this material is in its use as a photoanode for water oxidation. Illumination of a haematite electrode generates electron–hole pairs, and the holes can drive the oxidation of water at potentials that are more negative than the reversible oxygen potential. We can see this by running a linear sweep voltammogram while switching the light on and off. As Figure 8.10 shows, a photocurrent is generated by the illumination in the potential region where no, or very little, current flows in the dark. This is because the 4-electron oxygen evolution reaction summarized in equation (8.18) now involves transfer of electrons to the holes in the valence band and not electron tunnelling to the conduction band. The process is shown in the inset to Figure 8.10.

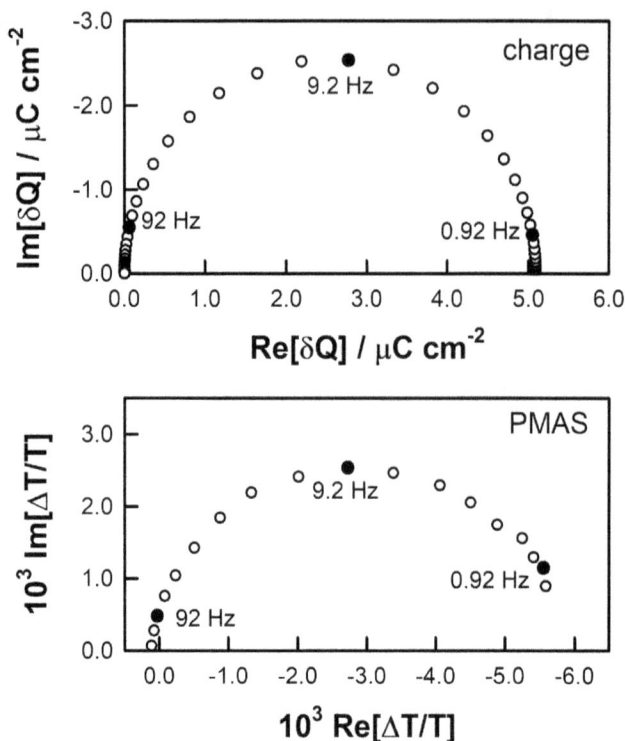

Figure 8.9. Nyquist plots comparing the modulated charge (top) and modulated absorbance (bottom) measured at 0.775 V vs. SCE (cf. Figure 8.7). The frequency dependence is clearly the same in both plots, showing that our model of the process is correct.

PMAS was able to detect the $Fe=O$ species formed during oxidation of water on haematite electrodes in the dark. Light-modulated absorbance spectroscopy (LMAS) can be used to detect species formed by illumination. LMAS is a small amplitude modulation technique like EIS, but it involves modulating the *light intensity* about a DC value rather than modulating the electrode potential. This can be achieved using the FRA (or lock-in amplifier) to drive a light-emitting diode while holding the potential constant with a potentiostat. Ideally, we would like to use a modulation of a few % superimposed on the DC component of the illumination, but in practice a higher amplitude modulation is usually necessary to obtain good spectra with acceptably low noise levels. The modulation of the light modulates the surface concentration of the intermediates we are trying to detect by absorbance (or reflectance) spectroscopy. The

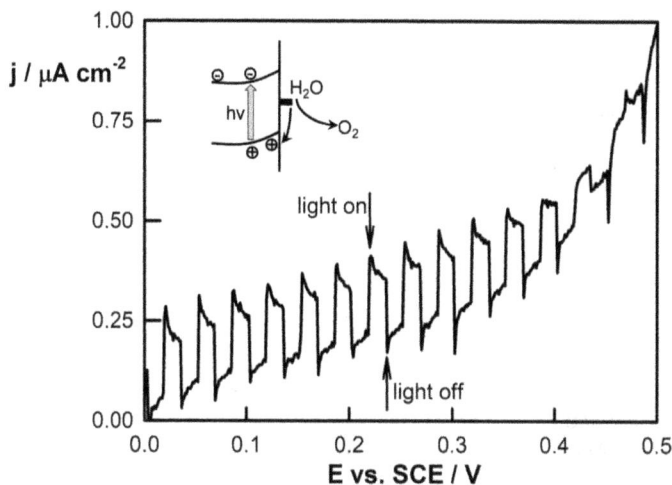

Figure 8.10. Voltammogram of nanocrystalline haematite electrode under chopped illumination from a UV LED (365 nm) showing the anodic photocurrent. The inset shows that illumination of the electrode generates electron–hole pairs, with the holes forming Fe=O and then finally oxygen in a 4-electron transfer process. LMAS can be used to identify intermediate species formed in the photoelectrochemical reaction.

first experimental problem we need to tackle is how to separate the excitation light from the lower-intensity monochromated light that used to measure the transmittance or reflectance spectrum. In the measurements on haematite, we used a UV LED (370 nm) as the modulated excitation source. A combination of slits and optical filters was then used to minimize the amount of 370 nm light entering the monochromator, which analyzes the transmitted white light at longer wavelengths than the excitation. The experimental setup is illustrated in Figure 8.11. The PMAS setup is similar except that the UV LED is absent and light intensity modulation is replaced by potential modulation.

The absorbance changes generated by illumination are smaller than those created by potential modulation, so longer integration times are required to reduce noise. Like the PMAS signal, the LMAS signal attenuates at higher frequencies, which is why a low (2.7 Hz) frequency is used for the potential modulation in PMAS and for the intensity modulation in LMAS. Figure 8.12 shows that LMAS can detect Fe=O on the electrode surface over a very wide potential range where no PMAS response is seen in the dark. The shape of the spectra is the same as seen in the PMAS spectrum, which tells us that the

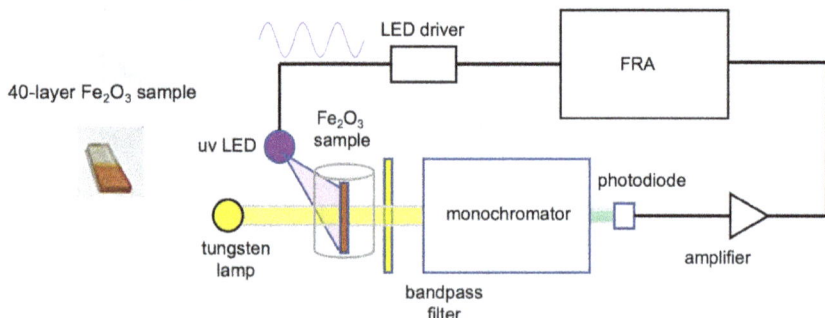

Figure 8.11. Experimental setup used for frequency-resolved LMAS on semi-transparent haematite layers deposited on FTO glass. A UV LED is used to photoexcite the haematite electrode to drive oxygen evolution. This LED is modulated by the signal from the FRA or lock-in amplifier. The transmittance of the semi-transparent sample is monitored using a white light source and a monochromator with a bandpass filter to eliminate any stray UV light. The output of the photodiode is fed back to the FRA (or to a lock-in amplifier for recording LMAS spectra).

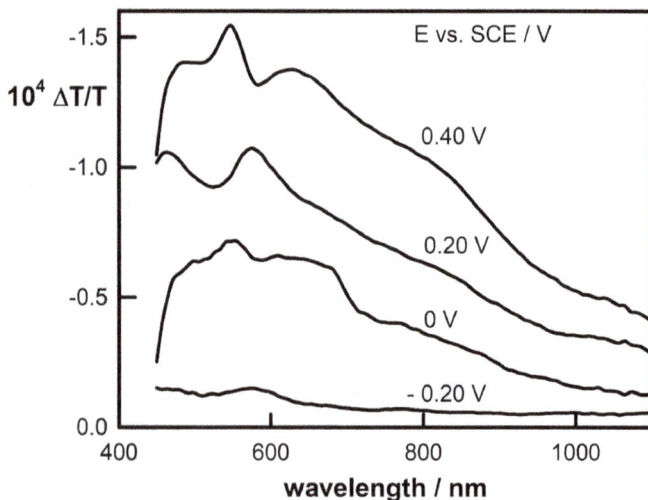

Figure 8.12. LMAS spectra obtained at an illuminated haematite electrode at different potentials. Modulated illumination 365 nm. Modulation frequency 2.7 Hz. Compare these spectra with the PMAS spectrum in Figure 8.7. They are the same, indicating that Fe=O is formed under illumination as well as in the dark during oxygen evolution.

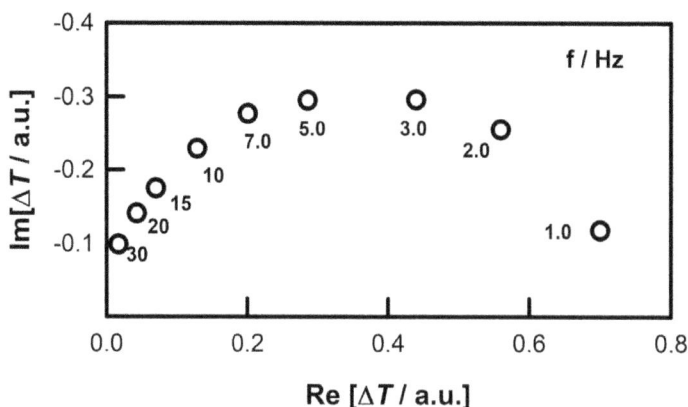

Figure 8.13. Nyquist plot of the LMAS response of a haematite electrode measured at a potential of 0.40 V vs. SCE and a wavelength of 600 nm. Frequencies as shown. The slow relaxation of the LMAS response indicates that the Fe=O state has a relatively long lifetime of 40 ms.

same intermediate is involved in both dark and light oxygen evolution reactions.

The frequency response of the LMAS signal is more difficult to obtain than for PMAS because the signal is around 50 times smaller. However, in this study, it was achieved using a lock-in amplifier and manual variation of the LED modulation frequency. An example of the results obtained is shown in Figure 8.13. The semi-circular Nyquist response is similar to the one obtained for PMAS (see Figure 8.9), indicating a long lifetime for the Fe=O intermediate. This lifetime can be estimated from $\tau_{Fe=O} = 1/2\pi f_{max} = 40$ ms.

References

Cummings, C. Y., Marken, F., Peter, L. M., Wijayantha, K. G. U. & Tahir, A. A. 2012. New insights into water splitting at mesoporous alpha-Fe_2O_3 films: A study by modulated transmittance and impedance spectroscopies. *Journal of the American Chemical Society*, 134, 1228–1234.

Kalaji, M. & Peter, L. M. 1991. Optical and electrical AC response of polyaniline films. *Journal of the Chemical Society-Faraday Transactions*, 87, 853–860.

Chapter 9

Intensity-Modulated Techniques: Application of IMPS and IMVS to Characterize Unconventional Solar Cells

9.1 Introduction

In Chapter 8, we saw that potential modulation can be replaced by modulation of light intensity when dealing with photoelectrochemical systems. This was the basis for light-modulated absorption spectroscopy (LMAS). Other intensity-modulated techniques have been developed to study solar cells as well as photoelectrochemical systems. In this chapter, we will look at two of the most important of these: *intensity-modulated photocurrent spectroscopy* (IMPS) and *intensity-modulated photovoltage spectroscopy* (IMVS). These involve small amplitude modulation of (usually monochromatic) light incident on the system being studied coupled with frequency-response analysis of the resulting current or voltage response. Related techniques are *light-modulated microwave reflectance spectroscopy* (LMMR) (Cass *et al.*, 2003; Dunn *et al.*, 2012) and *LMAS with near IR detection* (Franco *et al.*, 1999), both of which detect the perturbation of the electron concentration by the intensity-modulated illumination. This chapter explores the theoretical basis behind IMPS and IMVS and illustrates application of light-modulated techniques to dye-sensitized and perovskite solar cells.

9.2 Experimental Setup for IMPS

Intensity modulation of LED light sources is straightforward. One approach is to use one LED for the dc light source and a second, smaller, LED for the modulation. This small LED can be driven through a series resistor using the FRA to provide both a required dc level and the ac modulation. This setup is useful if high frequencies are required (up to 1 MHz), but a more usual alternative is to use an LED driver programmed from the FRA to produce both the dc level and the modulation of a larger LED. The downside is that the upper frequency limit of commercial LED drivers is usually less than 100 kHz. Turnkey solutions are provided by instruments such as the Solartron Modulab XM PhotoEchem. In this instrument, a high-speed reference photodiode is used to monitor the modulated light signal via a beam splitter. This avoids problems with attenuation and phase shift of the light signal due to the limitations of the LED driver. In order to maintain a constant % modulation while changing the illumination intensity, it is best to use calibrated neutral density filters rather than to adjust the input signals to the LED driver. A further requirement for high-frequency IMPS measurements is a fast potentiostat with minimal phase shift for current measurements. If a separate FRA and potentiostat are used, it is a good idea to use a low current sensitivity to minimize phase shifts in the current follower at high frequencies. To evaluate the upper frequency limit beyond which measurements become unreliable because of phase shift, the system can be tested using a fast photodiode to replace the solar cell.

The basic IMPS/IMVS setup is illustrated in Figure 9.1. For IMVS measurements at open circuit, the potentiostat used for IMPS is replaced by a high-impedance voltage follower. This can easily be built in-house using a suitable field-effect transistor (FET) operational amplifier powered by batteries (see Chapter 5).

The basic principle of IMPS and IMVS is illustrated by Figure 9.2, which shows two current–voltage plots for a typical solar cell calculated for two intensities that differ by 10%. The corresponding modulation of the short-circuit current is also 10%, but the percentage modulation of the open-circuit voltage is smaller due to the semilogarithmic dependence of V_{oc} on light intensity described by the diode equation (see Chapter 7).

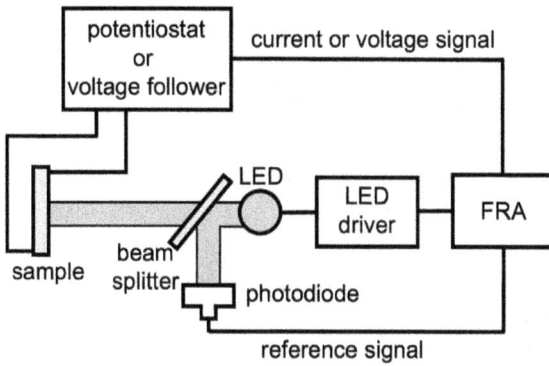

Figure 9.1. Basic experimental setup for IMPS and IMVS. The reference signal for the frequency-response analyzer is provided by the fast photodiode via a high bandwidth current amplifier (not shown).

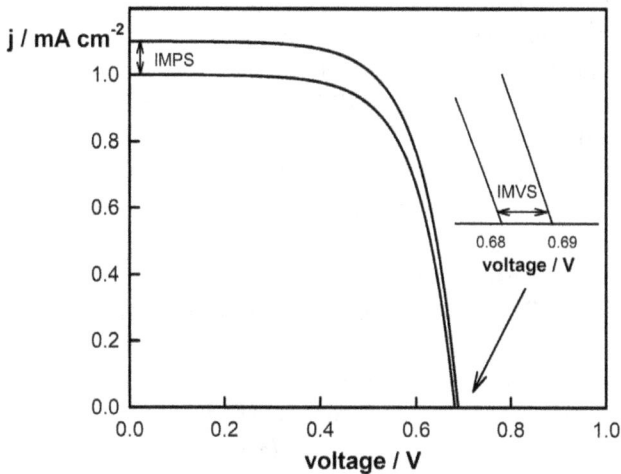

Figure 9.2. Current–voltage plots calculated for a typical solar cell for a 10% difference in illumination intensity. The corresponding changes in short-circuit current density and open-circuit voltage correspond to the low-frequency limits of the IMS and IMVS responses, respectively.

9.3 Application of IMPS and IMVS to Characterize Dye-Sensitized Solar Cells

Chapter 7 showed how PEIS can be used to characterize dye-sensitized solar cells. In PEIS measurements, the cell is illuminated with a dc light source, usually an LED, and the voltage across the cell is modulated. The impedance response contains information about the dye-sensitized TiO_2 photoelectrode as well as about the counter electrode kinetics and ion diffusion in the electrolyte. IMPS and IMVS are closely related techniques that involve modulating a light source by a few percent and measuring the resulting ac current (IMPS) or ac voltage (IMVS) response. IMPS measurements are usually made at short circuit using a potentiostat in two electrode mode, whereas the IMVS response at open circuit is measured using a high-impedance voltage follower to prevent distortion of the results by any leakage current (in principle, IMPS can be carried out at any fixed potential and IMVS at any fixed current, but this is rarely done). Both techniques contain the same information as PEIS, but, as we saw in Chapter 7, deconvolution of the electron transport resistance in PEIS measurement is sometimes difficult due to overlap with the counter electrode impedance.

The influence on the IMPS and IMVS responses of the other impedances in the DSSC has been analyzed in detail by Halme and co-workers (Halme, 2011; Halme *et al.*, 2008). However, for the purposes of this chapter, we will focus on the dye-sensitized TiO_2 electrode because its behaviour dominates the IMPS and IMVS responses measured at short circuit and open circuit, respectively.

The starting point for our discussion is the *generation/collection equation* for the mobile electrons in the DSSC. If we illuminate the cell through the conducting glass substrate ($x = 0$), the local rate of electron injection (electrons $cm^{-3} s^{-1}$) at any position x in the sensitized TiO_2 layer is determined by the product of the local light intensity (photons $cm^{-2} s^{-1}$) and the absorption coefficient $\alpha(\lambda)$ (cm^{-1}) at the wavelength λ used. Since the TiO_2 is transparent in the visible part of the spectrum, $\alpha(\lambda)$ is determined by the product of $\sigma_{opt}(\lambda)$, the optical cross section (cm^2) of the dye, and N_{dye}, the concentration (cm^{-3}) of dye molecules in the mesoporous TiO_2 layer: $\alpha(\lambda) = N_{dye}\sigma_{opt}(\lambda)$. N_{dye} is usually determined by desorbing the dye in an alkaline solution and measuring the optical absorbance of the resulting solution.

Electrons injected into the mesoporous TiO_2 from the photoexcited dye can either diffuse away from x or be lost by electron transfer to I_3^-. If we assume that the quantum efficiency of electron injection is 100%, the rate of change of electron concentration at any position x is given by

$$\frac{\partial n(x)}{\partial t} = \alpha I_0 e^{-\alpha x} + D_n \frac{\partial^2 n(x)}{\partial x^2} - \frac{n(x)}{\tau_n} \tag{9.1}$$

Looking at the right-hand side of equation (9.1), we recognize the first term as the generation (i.e., electron injection) term, which is given by the Beer–Lambert law. The second term is Fick's second law of diffusion, where D_n is the diffusion coefficient of electrons. The final term represents the loss of electrons by reaction with tri-iodide ions, the rate of which is assumed to be first order in electron concentration, so that the electron lifetime τ_n (s) is the inverse of the first-order rate constant (s^{-1}) for the reaction. The assumption of first-order kinetics supposes that the tri-iodide concentration is constant throughout the mesoporous TiO_2 layer. This will not be the case for a DSSC delivering high currents under strong illumination because the I_3^- ion concentration in the TiO_2 layer will be higher than the bulk value, and it will depend on position as a consequence of the ionic diffusion gradient. In this case, numerical solution of the generation collection problem becomes necessary.

If there is no barrier to the extraction of electrons at the conducting glass contact, the photocurrent will be *diffusion controlled*. There is a complication, however. Most of the electrons that are injected in the dye finish up in trap states where they cannot move. These electron traps are located at energies that lie in the forbidden gap of the TiO_2, and only electrons that are released by thermal excitation to the conduction band can contribute to the photocurrent. The variation of trap concentration with energy is exponential with a high concentration of shallow traps and a lower concentration of deep traps. The trap occupancy is determined by the *electron Fermi level*, which moves towards the conduction band energy as the light intensity is increased. As electrons move through the TiO_2 layer in the conduction band, they are repeatedly trapped and released, and the process is therefore referred to as *trap-controlled diffusion*. The end effect is that electron diffusion can be described by an 'effective' diffusion coefficient that increases as the electron Fermi level moves towards the conduction band with increasing illumination intensity. The same argument holds for the electron lifetime. Only free electrons can react with tri-iodide ions, and so τ_n is an 'effective' electron lifetime that decreases with increasing light intensity. We will not go into details here, but the good news is that if D_n and τ_n are measured *under the same conditions of trap occupancy*, L_n, the diffusion length of free electrons is still given by $L_n = \sqrt{D_n \tau_n}$.

Before we consider the transfer functions for IMPS and IMVS, we will look at the time (and hence frequency) scales involved. Starting with IMPS, consider the time t_n taken for electrons to diffuse from some position x within the TiO$_2$ film to the substrate ($x = 0$). It is given by

$$t_n(x) = \frac{x^2}{D_n} \tag{9.2}$$

At low light intensities, it can take seconds for electrons to move across the TiO$_2$ layer. This slow process is evident in the current response of a DSSC when light is switched on. The photocurrent rises slowly to a steady state and then decays equally slowly when the light is switched off. Since electrons are injected throughout the mesoporous layer, we expect the time delay, and hence the IMPS response, to represent an average of the diffusion times from different points weighted by the local generation (electron injection) rates. If the cell is illuminated from the substrate side with strongly absorbed light, the injected electrons will have a shorter distance to diffuse to the substrate contact than if the cell is illuminated from the electrolyte side. In the case of IMVS at open circuit, we are not extracting any electrons, so the ac component of the photovoltage should mainly reflect the decay of electron concentration due to reaction with tri-iodide ions. In addition, we expect to see some influence of diffusion if we use strongly absorbed light for IMVS because the electron concentration will tend to even out by diffusion.

Now let us turn to the exact solutions of the IMPS case. For IMPS, the photon flux is modulated by a few percent so that the generation term in equation (9.1) becomes time dependent.

$$\alpha I_0(t)e^{-\alpha x} = \alpha I_0 e^{-\alpha x}(1 + \delta e^{j\omega t}) \tag{9.3}$$

The solution of the generation/collection equation for small amplitude light modulation and diffusion-controlled electron extraction (for example, by using Laplace transforms) gives the *IMPS functions* for illumination via the substrate or via the electrolyte.

The *IMPS function* is defined as the complex ratio of the ac electron flux to the modulated incident photon flux. In terms of the photocurrent and photon flux, the IMPS function for illumination via the substrate takes

the dimensionless form

$$\Phi_{\text{IMPS,sub}}(\omega) = \frac{j_{\text{photo}}(\omega)}{q\delta I_0(\omega)} = \frac{\alpha}{\alpha+\gamma} \cdot \frac{\sinh(\gamma d) + \frac{\alpha}{\gamma-\alpha} \cdot (e^{-\alpha d} - e^{-\gamma d})}{\cosh(\gamma d)}$$

$$(9.4a)$$

where d is the thickness of the TiO$_2$ layer. Here the gamma term is a frequency-dependent complex quantity with units cm^{-1}.

$$\gamma = \sqrt{\frac{1}{D_n \tau_n} + \frac{\omega}{D_n}} \qquad (9.4b)$$

The dc solution is obtained by noting that for $\omega \to 0$, $\gamma d \to \frac{d}{L_n}$. For a 'good' DSSC, γd will be considerably less than 1, meaning that nearly all injected electrons arrive at the substrate. If $\gamma d > 1$, then a significant fraction of injected electrons will be lost during transit towards the substrate, and the cell efficiency will be low.

The IMPS function for illumination through the counter electrode and electrolyte is

$$\Phi_{\text{IMPS,el}}(\omega) = \frac{\alpha}{\alpha+\gamma} \cdot \frac{e^{-\alpha d}\sinh(\gamma d) + \frac{\alpha}{\gamma-\alpha} \cdot (e^{(\gamma-\alpha)d} - 1)}{\cosh(\gamma d)}$$

$$(9.5)$$

Figure 9.3 compares the shapes of the IMPS responses predicted for a 'good' DSSC ($L_n = 2.66d$) for the two illumination directions in the case of weakly absorbed light ($\alpha = 300$ cm^{-1}). This absorption coefficient is typical for DSSCs illuminated with red light close to the long wavelength onset of the absorption spectrum of commercially available ruthenium sensitizer dyes. Since the TiO$_2$ layer thickness in this case is 15 microns,

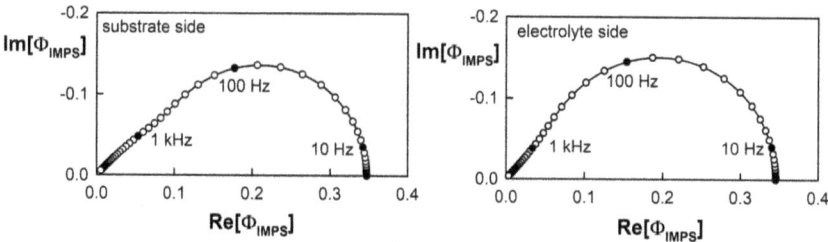

Figure 9.3. IMPS responses calculated from equations (9.4a) and (9.5) for illumination of a dye-sensitized solar cell with weakly absorbed light from the substrate and electrolyte sides, respectively. Thickness d of TiO$_2$ layer is 15 microns, absorption coefficient, α 300 cm^{-1}, 'effective' electron diffusion coefficient, 4.0×10^{-4} cm^2 s^{-1}, 'effective' electron lifetime, 4.0×10^{-2} s and electron diffusion length = $\sqrt{D_n \tau_n}$ = 40 microns.

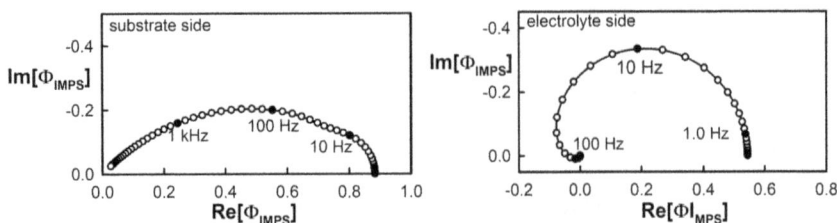

Figure 9.4. IMPS plots for substrate and electrolyte side illumination calculated for strongly absorbed light ($\alpha = 1500\,\mathrm{cm}^{-1}$). Thickness d of TiO$_2$ layer is 15 microns. 'Effective' electron diffusion coefficient, $4.0 \times 10^{-4}\,\mathrm{cm}^2\,\mathrm{s}^{-1}$, 'effective' electron lifetime, $4.0 \times 10^{-2}\,\mathrm{s}$. Electron diffusion length, 40 microns.

the electron injection rate drops by around half as the light penetrates the layer. Note how similar the shapes of the IMPS plots are to the *Bisquert distributed impedance* discussed in Chapter 7, with the high-frequency parts of Nyquist plots showing the –45° slope typical of a diffusion process. The correspondence is not surprising because the extended Bisquert#3 circuit element assumes uniform excitation, which is approximately the case for weakly absorbed light.

If strongly absorbed light is used for IMPS measurements, the consequences of inhomogeneous generation become apparent. Figure 9.4 contrasts the calculated IMPS responses for a much thicker TiO$_2$ layer (50 microns) and an absorption coefficient of 1500 cm^{-1}, which is a typical value for DSSCs illuminated with a green (530 nm) LED in the main part of the dye absorption spectrum. In this case the diffusion length is less than the layer thickness, and more electrons are lost when the electrode is illuminated from the electrolyte side than when light enters via the substrate. The shapes of the IMPS signatures are very different in the case of strongly absorbed light. The high-frequency response for electrolyte side illumination spirals into the origin because most electrons are injected close to the electrolyte side and have to diffuse across the whole TiO$_2$ layer to reach the substrate. This time delay results in a phase shift that increases with frequency, leading to the spiral response near the origin.

Normally, DSSCs are illuminated via the transparent conducting oxide-coated glass substrate to avoid losses arising from light absorption by the platinum-coated counter electrode and by tri-iodide ions in the electrolyte. However, DSSCs fabricated on metal substrates can only be illuminated from the counter electrode side. Examples of this configuration are DSSCs in which the mesoporous TiO$_2$ layer is replaced by an oriented layer of TiO$_2$

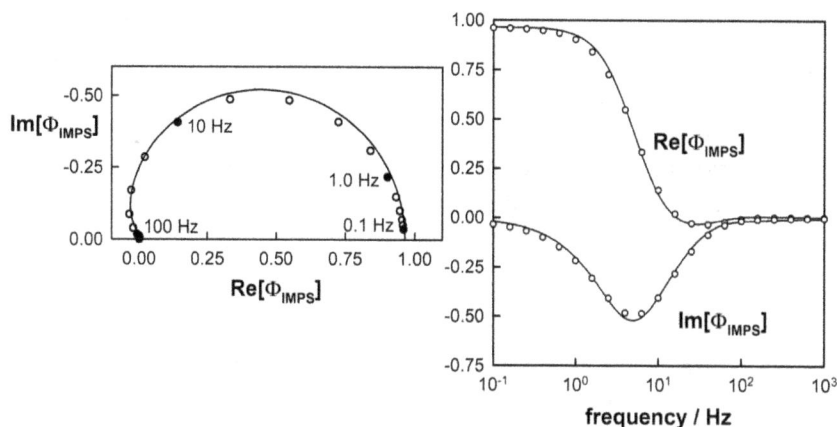

Figure 9.5. Nyquist and Bode plots of the experimental IMPS response of a DSSC fabricated using a 20-micron thick layer of oriented TiO$_2$ nanotubes. Illumination wavelength 530 nm. Points: experimental. Line: fit to equation (9.5). Fit parameters: $\alpha = 1740\,\text{cm}^{-1}$, $D_n = 5 \times 10^{-5}\,\text{cm}^2\,\text{s}^{-1}$, $\tau_n = 1.0\,\text{s}$. These values correspond to an electron diffusion length of around 70 microns, nearly twice the thickness of the nanotube layer.

nanotubes grown by anodization of a titanium metal substrate. The specific internal surface area of such a nanotube array is lower than the equivalent mesoporous layer, so the photoelectrode needs to be thicker in order to absorb most of the incident visible light. Figure 9.5 illustrates the Nyquist and Bode plots of the IMPS response of a DSSC fabricated in Bath by James Jennings using a 20-micron thick layer of oriented nanotubes grown in the Erlangen laboratory of Professor Patrik Schmuki by Dr Andrei Ghikov (James is now Senior Assistant Professor at the University of Brunei Darussalem). The absorption coefficient of the dye-coated nanotube layer measured by desorbing the dye and recording its absorption spectrum was found to be 1740 cm^{-1} at 530 nm, the LED wavelength used for the IMPS measurements. The IMPS response has been fitted to equation (9.4b) using this value of $\alpha(\lambda)$ to obtain the effective diffusion coefficient and electron lifetime. The quality of the fit is best seen in the Bode plot of the real and imaginary components of the IMPS function as a function of frequency (the incident light intensity was corrected for the transmittance of the platinized counter electrode). Further details of how the IMPS and IMVS data for the nanotube DSSCs are analyzed to obtain the electron diffusion length can be found in Jennings *et al.* (2008).

Now we will look at the IMVS of DSSCs at open circuit. If illumination is homogeneous, then the IMVS response is a semicircle with a maximum at the radial frequency $\omega_{\max} = 1/\tau_n$. This makes sense because the only way that the electron concentration can relax after injection is by transfer back to tri-iodide ions. The dependence of the DSSC photovoltage on light intensity can be described by the diode equation that we saw in Chapter 7:

$$V_{\text{oc}} = \frac{mk_BT}{q}\ln\left(\frac{j_{\text{sc}}}{j_s} + 1\right) \tag{9.6}$$

Here, j_{sc} is the short-circuit current density under illumination, j_s is the reverse saturation current density and m is the diode ideality factor. Since $j_{\text{sc}} \gg j_s$, equation (9.6) can be written in terms of the incident photon flux density I_0 and the external photocurrent quantum efficiency, EQE.

$$V_{\text{oc}} = -\frac{mk_BT}{q}\ln(j_s) + \frac{mk_BT}{q}\ln(I_0) = \text{const} + \frac{mk_BT}{q}\ln(I_0) \tag{9.7}$$

IMVS uses a small perturbation of the incident photon flux density $I_0(t) = I_0(1 + \delta e^{j\omega t})$, so if we note that $\ln(1 + \delta)$ can be approximated by δ for small perturbations, we find that the dc (i.e., $\omega \to 0$) perturbation of the open-circuit voltage is given by

$$\delta V_{\text{oc}} = \delta\frac{mk_BT}{q} \tag{9.8}$$

It follows that if we carry out IMVS measurements using weakly absorbed light and a constant value of δ while varying the dc intensity I_0, we should observe a series of semicircles of the same diameter in the Nyquist plots of the IMVS responses.

The exact IMVS transfer functions for inhomogeneous illumination via the substrate and via the electrolyte can be found in the same way as for IMPS using the boundary condition that the current through to substrate/TiO$_2$ interface is zero (this applies to open IMVS provided that a thin compact blocking layer of TiO$_2$ on the conducting glass is used to prevent electrons transferring with tri-iodide ions via the substrate). The expressions can be found, for example, in Halme *et al.* (2008).

Figure 9.6 shows a typical Nyquist plot of the IMVS response of a DSSC fabricated by Petra Cameron in our laboratory during her PhD studies (Petra is now Professor in the Department of Chemistry at Bath). These cells were fabricated with a thin compact TiO$_2$ layer under

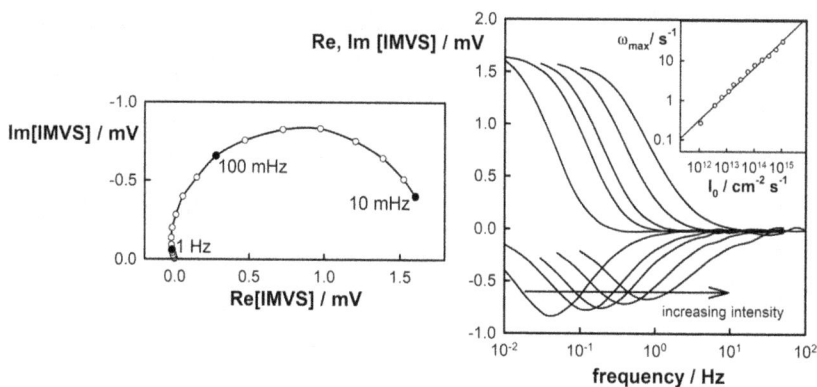

Figure 9.6. Left: Nyquist plot of the IMVS response of a DSSC illuminated via the substrate with strongly absorbed blue light (470 nm), showing the typical distortion of the semi-circular response at high frequencies. Right: Bode plots of real (upper) and imaginary (lower) components of the IMVS response (1.6 mV) of the same cell as a function of illumination intensity. The modulation depth δ was the same for all intensities. Note that the low-frequency (dc) limit of the IMVS response is independent of intensity, whereas ω_{max} shifts to higher frequencies as the intensity is increased, indicating that the 'effective' electron lifetime increases as electron trapping states in the mesoporous TiO$_2$ are progressively filled. The insert shows the intensity dependence of the frequency of the maximum in the imaginary component of the IMVS. TiO$_2$ layer thickness ca. 5 microns.

the mesoporous layer to prevent 'shunting' at open circuit by transfer of electrons to tri-iodide ions via the fluorine-doped tin oxide (FTO)-coated glass substrate. The IMVS measurements were made with strongly absorbed light (470 nm), so that the IMVS is not a perfect semicircle because there is some effect of electron diffusion at high frequencies. The Bode plots confirm that the low-frequency intercepts of the IMVS plots measured as a function of light intensity are constant as predicted by equation (9.8). The plots also show that the frequency of the maximum of the semicircle increases with light intensity. This is a consequence of the decrease in 'effective' electron lifetime with increasing occupancy of the electron trapping states in the TiO$_2$.

The preceding analysis of the IMVS response of DSSCs at open circuit assumes that the only route for the transfer of electrons to tri-iodide ions is via the mesoporous TiO$_2$. However, if electrons can pass to the conducting glass substrate and from there react with tri-iodide ions via areas that are in direct contact with the solution, a current will flow from the mesoporous electrode into the substrate at open circuit, so that the zero

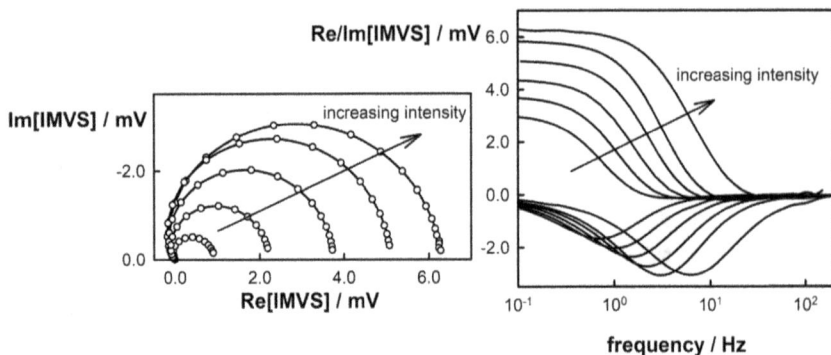

Figure 9.7. Nyquist and Bode plots of the IMVS response of a DSSC fabricated without the compact blocking layer of TiO_2 that is needed to prevent shunting of the cell via reduction of tri-iodide ions at exposed areas of the conducting glass substrate. Note that the low-frequency intercept is no longer independent of intensity but varies from ca. 3 mV to 6 mV because current flows to the electrolyte via the exposed FTO.

current boundary condition will no longer hold. This is the situation for DSSCs that are fabricated without a thin compact blocking layer of TiO_2. The effect of this shunting can be detected when IMVS measurements are made as a function of light intensity. Figure 9.7 illustrates the consequences of the shunting. The low-frequency intercept is now no longer constant but instead decreases as the light intensity is reduced. This is because the photovoltage is no longer correctly described by the diode equation as current leaks from the cell via the substrate. When this occurs, the values of electron lifetime obtained from the frequency of the maximum in the Nyquist plot are not reliable.

This conclusion is confirmed by measurements of the open-circuit voltage as a function of light intensity for cells with and without blocking layers, as shown in Figure 9.8. The cell with a blocking layer gives a linear plot of V_{oc} vs. $\log_{10}(I_0)$ that corresponds to a diode ideality factor of 1.4. The plot for the cell with no blocking layer is nonlinear with lower values of V_{oc} because current is flowing out of the DSSC via the uncoated substrate to reduce tri-iodide ions. At the highest light intensity, the open-circuit voltages coincide because the shunting current is small compared with the injection current. The fact that the slope of the semilogarithmic V_{oc} plot is much higher in the case of the DSSC with no blocking layer explains why the IMVS amplitude is also higher.

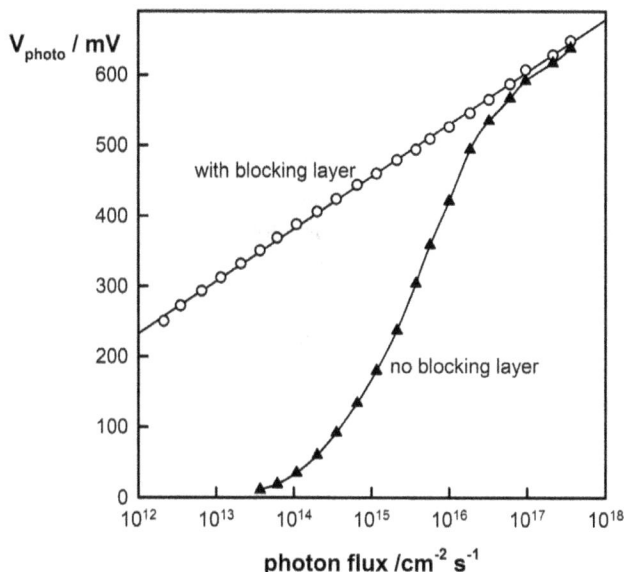

Figure 9.8. Plots of the open-circuit voltage as a function of incident photon flux density for DSSCs with and without compact blocking layers of TiO_2 The plots show that shunting of the cell via the uncoated substrate lowers the photovoltage. The effect is most pronounced at low light intensities because at higher light intensities the injection current is much larger than the current flowing via the substrate.

9.4 Relating the IMVS, IMPS and PEIS Responses of Perovskite Solar Cells

The PEIS response of perovskite solar cells was discussed in Chapter 7. Since the measurements were made at open circuit, it should be possible in principle to relate PEIS and IMPS/IMVS measurements. In the case of PEIS, the Fermi level in the device is modulated by application of an external ac voltage centred on V_{oc}, and an ac current flows through the external circuit. In the case of IMVS at open circuit, the Fermi level is modulated by changing the electron–hole pair generation rate, but no current flows through the external circuit. Since the impedance of the cell is defined as the complex ratio $\hat{Z} = \frac{\hat{V}}{\hat{i}}$, it should be possible to obtain the impedance from the ratio of the IMVS and IMPS responses. However, both responses must be measured under the same conditions, in other words at 'open circuit' (note that IMPS is normally performed at *short circuit*). Still, the meaning of 'open circuit' needs to be clarified. In the case of IMVS measurements made with a high-impedance voltage follower, the solar cell is truly at open circuit, and negligible current flows in the external circuit provided that the shunt resistance is high. By contrast, IMPS measurements have to be made at *the open-circuit potential* rather than 'at open circuit'. This means that the ac IMPS current flows through the external circuit. This current must flow through the series resistance, which in the case of the perovskite solar cell is mainly determined by the resistance of the conducting FTO-coated glass substrate. The same is true for PEIS measurements that are performed at the open-circuit potential. The relationship between the three different types of measurements is illustrated in Figure 9.9.

The magnitude of the attenuation by the series resistance of the IMPS signal measured at V_{oc} can be derived by considering the slope of the $j-V$ plot at V_{oc} from the diode equation.

$$j = j_{sc} - j_s e^{\frac{qV}{mk_BT}} \tag{9.9}$$

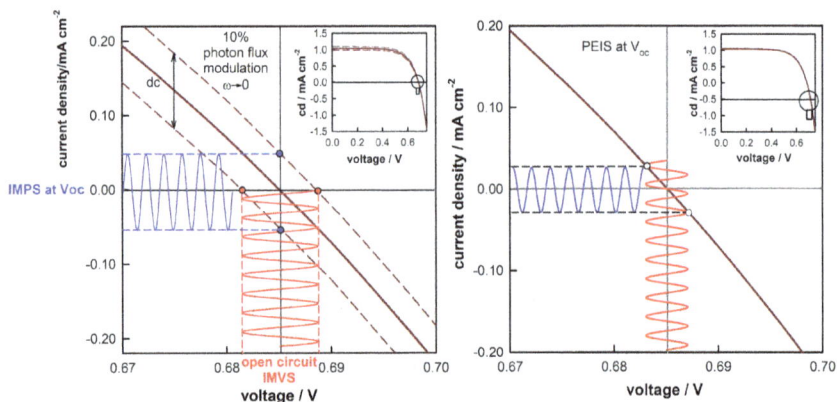

Figure 9.9. The left-hand figure shows the relationship between the IMVS and IMPS responses calculated from the current–voltage plots predicted by the equation for a solar cell for a 10% intensity modulation. In the case of IMVS at open circuit, the modulation creates a periodic change in V_{oc}, and no current flows in the external circuit. The amplitude of the IMPS corresponds to the distance between the two red points on the voltage axis. The IMPS amplitude corresponds to the vertical distance between the two blue points on the current axis. The right-hand plot shows the origin of the current response to potential modulation in the case of PEIS. The amplitude of the current response corresponds to the difference between the currents at the two points marked on the current–voltage plot.

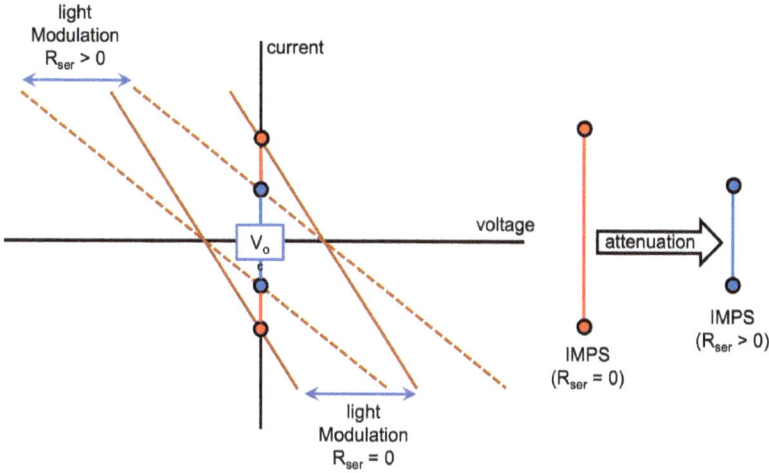

Figure 9.10. Diagram showing how the presence of a series resistance leads to attenuation of the IMPS response measured at the open-circuit voltage. The additional resistance decreases the slope of the current–voltage plots, and the figure shows how this affects the amplitude of the IMPS response.

Noting that $j = 0$ at V_{oc}, so that the terms on the right-hand side sum to zero, we find that the slope of the current–voltage plot at constant intensity is given by

$$\frac{\partial j}{\partial V} = \frac{q}{mk_BT} \cdot j_s e^{\frac{qV_{oc}}{mk_BT}} = \frac{q}{mk_BT} \cdot j_{sc} \tag{9.10}$$

The slope $\partial j/\partial V$ will be decreased by the series resistance R_{ser} as shown in Figure 9.10. This leads to attenuation of the IMPS current as shown. The attenuation factor in the low-frequency limit ($\omega \to 0$) is easily derived from the figure as

$$A_{\text{IMPS},V_{oc}} = \frac{1}{1 + R_{ser}\left(\frac{\partial j}{\partial V}\right)_{V_{oc}}} = \frac{1}{1 + \frac{q}{mk_BT} \cdot j_{sc}R_{ser}} \tag{9.11}$$

Taking some typical values: $j_{sc} = 1\,\text{mA}\,\text{cm}^{-2}$, $m = 1.4$, $R_{ser} = 20\,\Omega$, we find $A_{\text{IMPS},V_{oc}} = 0.64$, so the attenuation effect is significant. We can see this if we take some IMPS, IMVS and PEIS data measured by Adam Pockett on the same planar perovskite cell in Professor Petra Cameron's laboratory in Bath. The Nyquist plots of the IMVS and IMPS responses are shown in Figure 9.11 together with the measured PEIS response and

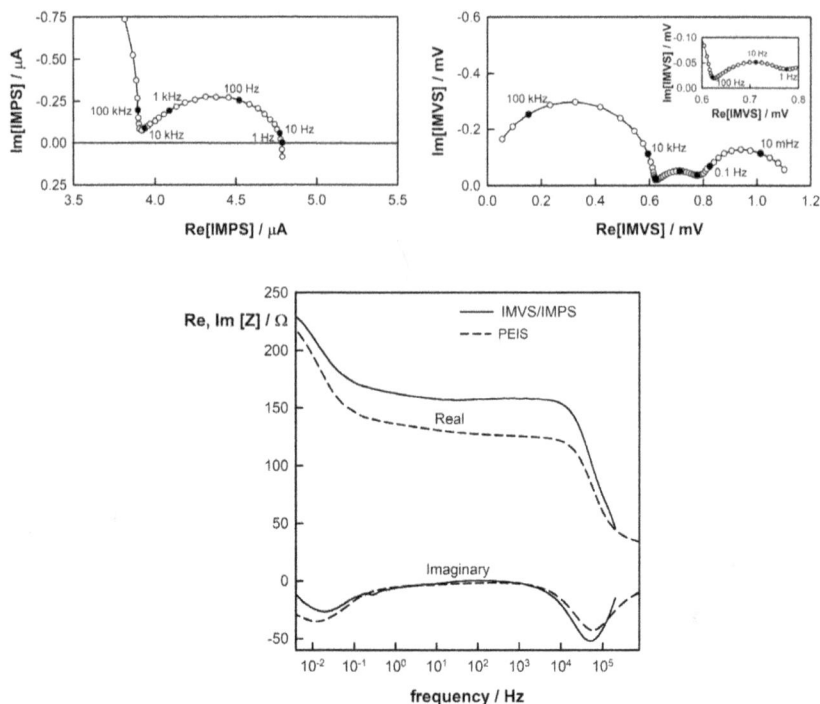

Figure 9.11.   Top: Measured IMVS and IMPS responses of a planar perovskite solar cell. Bottom: Measured PEIS response compared with the PEIS response predicted from the ratio of the IMVS and IMPS responses without consideration of the attenuation of the IMPS response by the series resistance. Data kindly provided by Dr Adam Pockett and Professor Petra Cameron. Note that the predicted impedance is higher than the measured impedance due to the neglect of the attenuation factor.

the impedance response calculated from the IMPS/IMVS ratio without considering attenuation of the IMPS signal by series resistance.

Interpretation of the IMVS, IMPS and PEIS responses shown in Figure 9.11 is complicated by the fact that the metal halide perovskites are mixed electronic/electronic conductors. The high-frequency semicircle seen in both the IMVS and PEIS is expected for a normal solar cell. It arises from the parallel combination of the recombination resistance and the geometric capacitance (the geometric capacitance is larger than the chemical capacitance for planar thin-layer perovskite solar cells – see Chapter 7). The low-frequency response, on the other hand, is associated with the interaction of mobile charged vacancies with electrons and holes as discussed in Chapter 7.

References

Cass, M. J., Duffy, N. W., Peter, L. M., Pennock, S. R., Ushiroda, S. & Walker, A. B. 2003. Microwave reflectance studies of photoelectrochemical kinetics at semiconductor electrodes. 1. Steady-state, transient, and periodic responses. *Journal of Physical Chemistry B*, 107, 5857–5863.

Dunn, H. K., Peter, L. M., Bingham, S. J., Maluta, E. & Walker, A. B. 2012. In situ detection of free and trapped electrons in dye-sensitized solar cells by photo-induced microwave reflectance measurements. *Journal of Physical Chemistry C*, 116, 22063–22072.

Franco, G., Gehring, J., Peter, L. M., Ponomarev, E. A. & Uhlendorf, I. 1999. Frequency-resolved optical detection of photoinjected electrons in dye-sensitized nanocrystalline photovoltaic cells. *Journal of Physical Chemistry B*, 103, 692–698.

Halme, J. 2011. Linking optical and electrical small amplitude perturbation techniques for dynamic performance characterization of dye solar cells. *Physical Chemistry Chemical Physics*, 13, 12435–12446.

Halme, J., Miettunen, K. & Lund, P. 2008. Effect of nonuniform generation and inefficient collection of electrons on the dynamic photocurrent and photovoltage response of nanostructured photoelectrodes. *Journal of Physical Chemistry C*, 112, 20491–20504.

Jennings, J. R., Ghicov, A., Peter, L. M., Schmuki, P. & Walker, A. B. 2008. Dye-sensitized solar cells based on oriented TiO_2 nanotube arrays: Transport, trapping, and transfer of electrons. *Journal of the American Chemical Society*, 130, 13364–13372.

Chapter 10

Applications of IMPS and PEIS to Study Photoelectrode Kinetics

10.1 Introduction

In this final chapter, we look at how frequency-resolved techniques such as intensity-modulated photocurrent spectroscopy (IMPS) and photoelectrochemical impedance spectroscopy (PEIS) can be used to study the kinetics of light-driven reactions at semiconductor electrodes. The most important reactions are those involved in the photoelectrolysis of water to produce hydrogen and oxygen (light-driven water splitting). These reactions are *multistep electron/proton transfer reactions* that involve surface bound intermediates. The reactions are remarkably slow on semiconductor electrodes in the absence of electrocatalysts, and this makes it difficult to achieve high solar-to-hydrogen conversion efficiencies (STH efficiencies) because the desired interfacial reactions must compete with recombination reactions. The mechanistic complexity of photoelectrochemical oxygen evolution on haematite (α-Fe_2O_3) was already discussed in Chapter 8. There we saw that a likely mechanism of the four-electron, four-proton oxidation of water on haematite in alkaline solution in the dark involves several surface bound species. Under illumination, the same reaction involves photogenerated holes in the equivalent reaction scheme

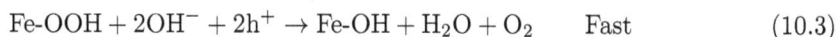

$$\text{Fe-O}^- + \text{h}^+ \rightleftharpoons \text{Fe=O} \qquad \text{First hole capture} \qquad (10.1)$$

$$\text{Fe=O} + \text{OH}^- + \text{h}^+ \rightarrow \text{FeOOH} \qquad \text{Slow rate-determining step} \qquad (10.2)$$

$$\text{Fe-OOH} + 2\text{OH}^- + 2\text{h}^+ \rightarrow \text{Fe-OH} + \text{H}_2\text{O} + \text{O}_2 \qquad \text{Fast} \qquad (10.3)$$

The details of this reaction scheme are still being explored at the time of writing this chapter, but we will look at how IMPS and PEIS have been used to investigate the kinetics of the photoelectrochemical oxygen evolution reaction (POER) on haematite photoanodes.

The other half of the water splitting reaction is of course photoelectrochemical hydrogen evolution (PHER). The first (rate-determining) step in the two-electron, two-proton photo-reduction of water on p-type semiconductors such as p-GaP and Si appears to involve the reaction of bound hydrogen atoms on hydrogen-terminated surfaces of the semiconductors with electrons and protons to form dihydrogen, followed by replacement of the abstracted hydrogen atom by electron transfer to a second proton. For example, on a p-Si(111) surface, the process may take place as follows:

$$\equiv \text{Si-H} + \text{H}^{+\bullet} + \text{e}^- \rightarrow \equiv \text{Si} + \text{H}_2 \qquad \text{Slow rate-determining step} \qquad (10.4)$$

$$\equiv \text{Si}^\bullet + \text{H}^+ + \text{e}^- \rightarrow \equiv \text{Si-H} \qquad \text{Fast} \qquad (10.5)$$

Here, $\equiv$Si-H represents the termination of each silicon atom in the (111) crystal surface by one hydrogen, and $\equiv$Si$^\bullet$ represents the unterminated silicon atom in the surface formed by the first step in the reaction sequence. The hydrogen formed by the first step is known to dissociate and penetrate into the p-type silicon crystal lattice, taking up interstitial positions associated with acceptor dopants such as boron. This absorption process leads to the formation of 'near-surface states' that act as recombination centres.

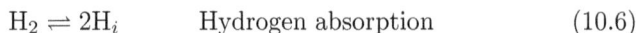

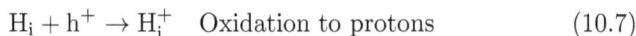

$$\text{H}_2 \rightleftharpoons 2\text{H}_i \qquad \text{Hydrogen absorption} \qquad (10.6)$$

$$\text{H}_i + \text{h}^+ \rightarrow \text{H}_i^+ \qquad \text{Oxidation to protons} \qquad (10.7)$$

Here, H_i represents a hydrogen atom at an interstitial position. The protons formed in the near-surface region in the second step are highly mobile and can transfer to the solution. The two-step reaction sequence is therefore equivalent to the electron–hole recombination.

Unravelling the mechanisms and kinetics of the POER and PHER represents a major ongoing challenge, but before we address this problem, we need to understand some of the basic concepts of semiconductor photoelectrochemistry.

10.2 A Crash Course in Semiconductor Photoelectrochemistry

Semiconductor photoelectrochemistry is a broad topic, and this section only provides a highly condensed account of the main points that are essential to understand how IMPS and PEIS are used to study photoelectrochemical reactions such as those involved in light-driven water splitting to generate 'green hydrogen'. Several books and reviews are available that provide an in-depth survey of the field (e.g., Memming 2015; Sato 1998).

An n-type semiconductor is one that has been doped (intentionally or unintentionally) with an electron donor. As a consequence, the electron concentration in the semiconductor is much higher than the hole concentration. In the neutral semiconductor, the ionized electron donor atoms that are fixed in the crystal lattice carry a positive charge that balances the negative charge of the electrons that originate from the donors. When the semiconductor is contacted by a suitable redox electrolyte in the dark, electrons flow from the semiconductor to the redox species until the Fermi levels of the semiconductor and redox electrolyte are equal at equilibrium. This removal of electrons in the equilibration process leaves behind the ionized donors as a *space charge region* (SCR) that generates an electrical field that varies linearly across the space charge region. As a consequence, the potential energy of electrons and holes depends on distance in such a way that the bands are bent in the parabolic shape illustrated in Figure 10.1. In the absence of a 'fast' redox system, the band bending can be controlled by altering the electrode potential, i.e., the electrode is polarizable. This is the situation in the case of photoelectrochemical water splitting since the redox reactions are too slow to establish electronic equilibrium.

The same concepts apply to a p-type electrode, except that in this case we are dealing with electron acceptors that have a negative charge when they have removed electrons from the valence band to create holes. Since the space charge is negative in this case, the sign of the electrical field and the direction of band bending are reversed. In the following discussion, it is assumed that we are dealing with an n-type semiconductor electrode (i.e., a photoanode).

The width of the space charge region, W_{sc}, depends on the doping concentration N_d and on the potential drop $\Delta\phi_{sc}$ between the bulk of the

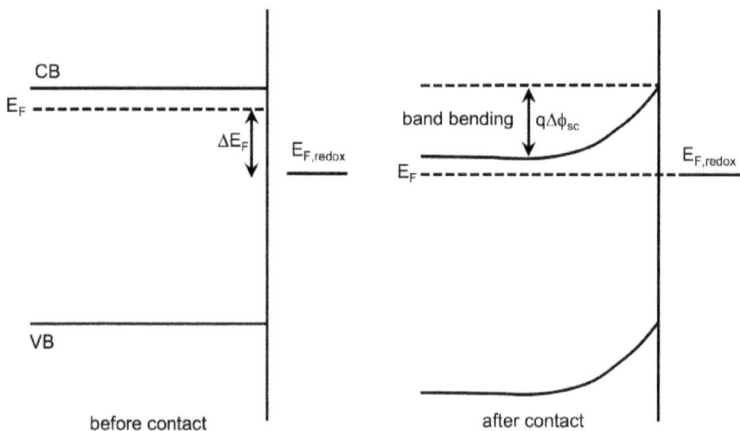

Figure 10.1. Equilibration of an n-type semiconductor electrode with a redox system with a lower Fermi energy leads to creation of a space charge region that results in band bending. The band bending $q\Delta\phi_{sc}$ is equal to the initial difference in Fermi energies of the electrode and redox electrolyte. In the absence of a fast redox system, the band bending can be controlled by altering the potential. This is the situation for photoelectrodes in contact with inert aqueous electrolytes. The potential at which the space charge vanishes along with the band bending is referred to as the flat band potential.

semiconductor and the surface (i.e., on the band bending). It also depends on the relative permittivity (dielectric constant) ε of the semiconductor as shown by equation (10.8). ε_0 is the permittivity of free space.

$$W_{sc} = \left(\frac{2\Delta\phi_{sc}\varepsilon\varepsilon_0}{qN_d} \right)^{1/2} \tag{10.8}$$

Here, $\Delta\phi_{sc}$ is the potential difference across the space charge region, and N_d is the doping density. The immobile positive charge in the space charge region of an n-type photoelectrode is balanced by a net negative charge in the electrolyte brought about by the relative displacement of anions and cations in the electrical double layer. This arrangement of charges defines the *space charge capacitance*, C_{sc}, which is given by the *Mott Schottky equation*.

$$\frac{1}{C_{sc}^2} = \frac{2}{qN_d\varepsilon\varepsilon_0} \left(E - E_{fb} - \frac{k_BT}{q} \right) \tag{10.9}$$

Here, E is the electrode potential, and E_{fb} is the *flatband potential*, the potential at which the space charge region disappears, and the band bending is zero.

When the n-type semiconductor absorbs photons with an energy greater than its bandgap, electron–hole pairs are created. The holes vacancies in the valence band are the *minority carriers*, and the electrons are the *majority carriers*. In the neutral bulk of the semiconductor, recombination of holes is fast because there is a high concentration of electrons. By contrast, the electron concentration is much lower in the space charge region than in the bulk semiconductor so that holes have a high chance of reaching the interface, where they can receive electrons from species in the electrolyte, resulting in an oxidation reaction. Holes created in the neutral region near the edge of the space charge region also have a chance of diffusing into the space charge region that depends on their diffusion length, L_p, which is the square root of the product of their lifetime and diffusion coefficient: $L_p = \sqrt{\tau_p D_p}$. In 1959, this generation/collection problem was solved for the steady-state and for high-frequency modulation conditions by Wolfgang Gärtner of the US Signals Research and Development Laboratory (Gärtner, 1959). The dependence of hole current on the width of the space charge region (and hence on band bending or electrode potential) is given by the *Gärtner equation*

$$j_G = qI_0 \frac{e^{-\alpha W_{\mathrm{sc}}}}{1 + \alpha L_p} \tag{10.10}$$

The derivation of the Gärtner equation assumes that all holes reaching the surface are transferred across the interface (in our case, this means that they accept electrons from electrolyte species). However, holes may instead recombine with conduction band electrons or with electrons trapped in surface electronic states located in the band gap – so-called *surface states*.

In summary then, electrode reactions at illuminated semiconductor electrodes involve the following steps: (1) light absorption to create electron–hole pairs; (2) separation of electrons and holes driven by the gradients of the respective *quasi-Fermi levels* (see Appendix 7.2 in Chapter 7); (3) electron transfer reactions involving the respective minority carriers (holes for an n-type photoanode and electrons for a p-type photocathode); (4) recombination of electrons and holes in the bulk and at the surface. A simplified sketch illustrating these processes for an n-type semiconductor electrode is shown in Figure 10.2.

The simplified scheme in Figure 10.2 allows us to define the fraction of holes that successfully react at the surface and avoid recombination. We call

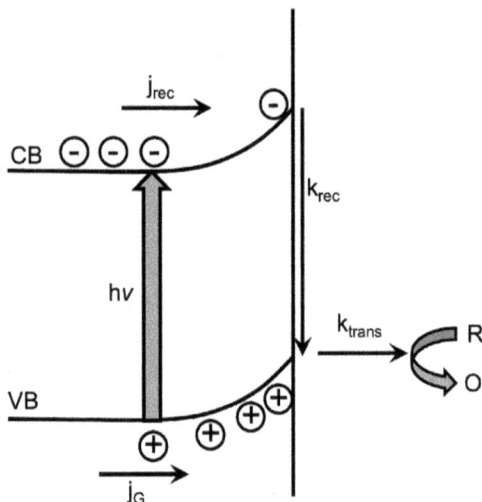

Figure 10.2. A simplified picture of the processes taking place at an illuminated n-type photoanode. Photogenerated electron–hole pairs are separated, with the minority carrier (in this case, the holes) moving to the interface, where they can drive oxidation reactions. Electrons and holes can also recombine in the bulk and at the surface. Electron transfer and surface recombination are described by first-order rate constants k_{trans} and k_{rec}. These rate constants represent a simplification in the case of multistep photoelectrode reactions such as those involved in water splitting.

this the *transfer efficiency*, η_{trans}

$$\eta_{trans} = \frac{k_{trans}}{k_{trans} + k_{rec}} \tag{10.11}$$

The corresponding *dc photocurrent*, j_{photo}, measured by a potentiostat will be given by

$$j_{photo} = j_G \eta_{trans} \tag{10.12}$$

The recombination rate of holes at the surface depends on the electron and hole concentrations at the surface, n_{surf}. In the absence of surface states, this electron concentration is determined by the band bending, i.e., by the voltage drop across the space charge region. If, on the other hand, electrons are trapped in surface states, their concentration will be determined by the density of states distribution of the surface states and the electron Fermi level at the surface.

For simplicity, we will leave aside the complications associated with surface states and assume that n_{surf} is not changed significantly by illumination (i.e., the electron Fermi level is not perturbed). We can then define a pseudo-first-order rate constant k_{rec} that is therefore potential dependent as shown by equation (10.13).

$$k_{rec} = k'_{rec} n_{surf} = k'_{rec} n_{bulk} e^{-\frac{\beta q \Delta \phi_{sc}}{k_B T}} = k'_{rec} N_d e^{-\frac{\beta q \Delta \phi_{sc}}{k_B T}} \qquad (10.13)$$

Here, n_{bulk} is the concentration of electrons in the bulk of the semiconductor, which is the same as the doping concentration N_d, and k'_{rec} is a second-order rate constant. In the ideal case, the term $\beta = 1$, but experimentally we generally find that $\beta < 1$. It should be noted that the assumption of pseudo-first-order recombination kinetics is likely to break down under conditions where electrons are no longer present in large excess. This will be the case for high band banding and slow transfer kinetics, where the hole concentration at the surface can become large.

10.3 A Closer Look at the Phenomenological Rate Constants k_{trans} and k_{rec}

It is important to realize that the scheme shown in Figure 10.2 is a 'phenomenological' kinetic description of the processes taking place at a photoelectrode. This means that linking it to a description of processes taking place at a molecular level requires a mechanistic model. To obtain a deeper understanding of the significance of the phenomenological electron transfer and recombination rate constants, we can examine, for example, the mechanism proposed for photoelectrochemical oxygen evolution on haematite outlined in equations (10.1)–(10.3). In total, four holes are transferred for each dioxygen molecule produced, so it is not clear what we mean by k_{trans}, the rate constant for charge transfer. The concept of recombination also needs to be clarified. In the simple scheme, recombination refers to electron–hole recombination, but as we have seen previously, 'recombination' may actually occur indirectly via the surface-bound species formed as intermediates in the overall reaction. A modification of the scheme in equations (10.1)–(10.3) shows how 'recombination' also refers to these indirect routes.

$$\text{Fe-O}^- + \text{h}^+ \rightleftharpoons \text{Fe=O} \quad \text{First hole capture} \qquad (10.14a)$$

$$\text{Fe=O} + \text{e}^- \rightleftharpoons \text{Fe-O}^- \quad \text{Back reaction with cb electrons} \qquad (10.14b)$$

Fe=O + OH$^-$ + h$^+$ → FeOOH   Slow rate-determining step   (10.15a)

FeOOH + e$^-$ → Fe=O + OH$^-$ Back reaction with cb electrons   (10.15b)

Fe-OOH + 2OH$^-$ + 2h$^+$ → Fe-OH + H$_2$O + O$_2$ + 2e$^-$ Fast   (10.16)

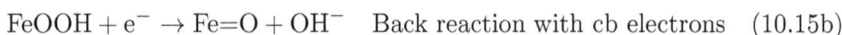

The scheme shows that the slow rate determining step Fe=O + h$^+$ → FeOOH must compete with the reverse electron capture step Fe=O + e$^-$ → Fe-O$^-$. The subsequent reactions (10.15a)–(10.16) leading to the formation of oxygen are likely to be much faster than electron capture. In this case, we can conclude that the net effect of reaction (10.14a) followed by (10.14b) is the recombination of an electron and a hole. It is possible to derive the IMPS and PEIS responses for this specific mechanistic scheme, but we will not do so here since it lies outside the scope of this book. Instead, we will continue with the original phenomenological approach pioneered by my collaborator Evgueni Ponomarev, which has proved useful in the analysis of experimental data for photoelectrochemical oxygen evolution on n-type photoanodes and photoelectrochemical hydrogen evolution of p-type photocathodes.

10.4 The IMPS Transfer Function Φ_{IMPS}

IMPS involves superimposing a small periodic modulation on a dc illumination level, I_0. The incident photon flux $I(t)$ is described by the function

$$I(t) = I_o(1 + \delta e^{j\omega t}) \tag{10.17}$$

where δ is typically of the order of a few %. The modulation of the incident photon flux creates a periodic change in the hole current, j_G, predicted by the Gärtner equation (equation (10.10)). For the frequency range normally used in IMPS (<100 kHz), we can assume that the carrier transit time effects considered by Gärtner for the very high-frequency response are negligible, so that modulated hole current will remain in phase with the illumination, i.e.,

$$\delta j_G(t) = |\delta j_G| e^{j\omega t} \tag{10.18}$$

This modulation of the hole current leads to a time-dependence of the hole charge at surface, $q\delta p(t)$. The rate of change of the surface hole charge is determined by the rates of arrival and removal of holes by interfacial electron transfer and recombination as follows:

$$q\frac{d\delta p(t)}{dt} = \delta j_G(t) - k_{trans}q\delta p(t) - k_{rec}q\delta p(t) \tag{10.19}$$

This surface charging process corresponds to charging the space charge capacitance, which is much smaller than the Helmholtz capacitance (unless the semiconductor is very highly doped), so that we can write equation (10.19) in the equivalent form

$$j_{charging} = j_{hole} - j_{transfer} + j_{recombination} \tag{10.20}$$

Note that both the hole current and the transfer current have a positive sign, whereas the recombination current has a negative sign because it corresponds to the flow of negatively charged electrons into the interface.

The photocurrent measured in the external circuit is

$$j_{photo} = j_{hole} + j_{recombination} = j_{charging} + j_{transfer} \tag{10.21}$$

where the hole current is positive, and the recombination current is negative. It is important to realize that an ac photocurrent will be measured under modulated illumination, *even if no charge transfer occurs*. In this case, the photocurrent just corresponds to charge and discharge of the space charge capacitance. Under dc conditions, by contrast, the photocurrent will be zero in the absence of charge transfer because the charging current is absent. This is exactly the kind of behaviour we expect for a capacitor.

Since $\delta j_G(t) = |\delta j_G|e^{j\omega t}$, we can assume that the small modulation of the surface hole charge will also be sinusoidal with some phase shift ϕ, allowing us to write equation (10.19) as

$$q\frac{d}{dt}\left(|\delta p|e^{(j\omega t+\phi)}\right) = |\delta j_G|e^{j\omega t} - (k_{trans} + k_{rec})q|\delta p|e^{(j\omega t+\phi)} \tag{10.22}$$

Noting that $\frac{d}{dt}\left(e^{(j\omega t+\phi)}\right) = j\omega e^{(j\omega t+\phi)}$, we find that

$$q\left(|\delta p|j\omega e^{(j\omega t+\phi)}\right) = |\delta j_G|e^{j\omega t} - (k_{trans} + k_{rec})q|\delta p|e^{(j\omega t+\phi)} \tag{10.23}$$

The modulated surface hole charge is therefore

$$q\left(|\delta p|e^{(j\omega t+\phi)}\right) = \frac{|\delta j_G|e^{j\omega t}}{(k_{trans} + k_{rec} + j\omega)} \tag{10.24}$$

Finally, the modulated photocurrent in the external circuit is given by the sum of the hole current into the surface and the electron current into the surface due to recombination.

$$\delta j_{\text{photo}}(t) = \delta j_G - q k_{\text{rec}} \delta p(t) = |\delta j_G| e^{j\omega t} - \frac{k_{\text{rec}} |\delta j_G| e^{j\omega t}}{(k_{\text{trans}} + k_{\text{rec}} + j\omega)} \quad (10.25)$$

Now we can define the IMPS function, Φ_{IMPS}, as the ac equivalent of the transfer efficiency, η_{trans}, which was defined in equation (10.11)

$$\Phi_{\text{IMPS}} = \frac{\delta j_{\text{photo}}(t)}{\delta j_G(t)} = \frac{1}{|\delta j_G| e^{j\omega t}} \left(|\delta j_G| e^{j\omega t} - \frac{k_{\text{rec}} |\delta j_G| e^{j\omega t}}{(k_{\text{trans}} + k_{\text{rec}} + j\omega)} \right)$$
$$(10.26)$$

which simplifies to

$$\Phi_{\text{IMPS}} = 1 - \frac{k_{\text{rec}}}{(k_{\text{trans}} + k_{\text{rec}} + j\omega)} = \frac{k_{\text{trans}} + j\omega}{k_{\text{trans}} + k_{\text{rec}} + j\omega} \quad (10.27)$$

Since Φ_{IMPS} is a complex quantity, we can obtain its real and imaginary components in the usual way using the complex conjugate of the denominator.

$$\text{Re}\Phi_{\text{IMPS}} = \frac{k_{\text{trans}}(k_{\text{trans}} + k_{\text{rec}}) + \omega^2}{(k_{\text{trans}} + k_{\text{rec}})^2 + \omega^2} \quad (10.28a)$$

$$\text{Im}\Phi_{\text{IMPS}} = \frac{k_{\text{rec}}\omega}{(k_{\text{trans}} + k_{\text{rec}})^2 + \omega^2} \quad (10.28b)$$

In the low-frequency limit,

$$\text{Re}\Phi_{\text{IMPS},\omega \longrightarrow 0} = \frac{k_{\text{trans}}}{k_{\text{trans}} + k_{\text{rec}}} \quad (10.29a)$$

$$\text{Im}\Phi_{\text{IMPS},\omega \longrightarrow 0} = 0 \quad (10.29b)$$

In the high-frequency limit,

$$\text{Re}\Phi_{\text{IMPS},\omega \longrightarrow \infty} \longrightarrow 1 \quad (10.30a)$$

$$\text{Im}\Phi_{\text{IMPS},\omega \longrightarrow \infty} \longrightarrow 0 \quad (10.30b)$$

The IMPS function describes a semicircle as shown in Figure 10.3. The figure also shows how the in-phase hole current and the out-of-phase recombination current sum up to give the real and imaginary components of the IMPS function at any particular frequency.

Differentiating equation (10.28b) with respect to ω to find the maximum in $\text{Im}\Phi_{\text{IMPS}}$ reveals that the maximum of the semicircle is located at a

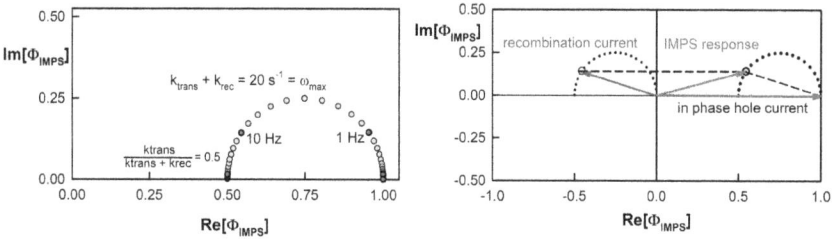

Figure 10.3. Left: The Nyquist plot of the IMPS function is a semicircle with a maximum located at $\omega_{max} = (k_{trans} + k_{rec})$ and a low-frequency intercept at $k_{trans}/(k_{trans} + k_{rec})$, which is equal to the dc transfer efficiency η_{trans}. In the case shown, $k_{trans} = k_{rec} = 10\,\mathrm{s}^{-1}$. Right: At any particular frequency, addition of the frequency-independent in-phase hole current vector (red) and the frequency-dependent out-of-phase recombination current (blue) gives the IMPS vector (green). Note that the recombination current vectors for different frequencies also trace out a semicircle.

radial frequency $\omega_{max} = (k_{trans} + k_{rec})$. It follows that k_{trans} and k_{rec} can be determined from ω_{max} and $\mathrm{Re}\Phi_{IMPS,\omega \rightarrow 0}$.

10.5 The Influence of Capacitance and Resistance on the IMPS Response

The IMPS function describes the frequency dependence of the photocurrent that is generated by the processes in the semiconductor. In practice, this current will have to flow through a series resistor to be measured in the external circuit. This resistance corresponds to the sum of the contact and electrolyte resistances if we are using a potentiostat. The generation of a voltage drop across the series resistance will cause the electrode potential to deviate from the applied value, which in turn leads to charging of the space charge and Helmholtz capacitances. This process can be considered within the framework of Kirchhoff's voltage and current laws. The result is that the IMPS signal is attenuated at higher frequencies. The IMPS function becomes

$$\Phi_{IMPS} = \frac{k_{trans} + j\omega(C/C_{sc})}{k_{trans} + k_{rec} + j\omega} \left(\frac{1}{1 + j\omega RC} \right) \tag{10.31a}$$

where

$$C = \frac{C_{sc}C_H}{C_{sc} + C_H} \tag{10.31b}$$

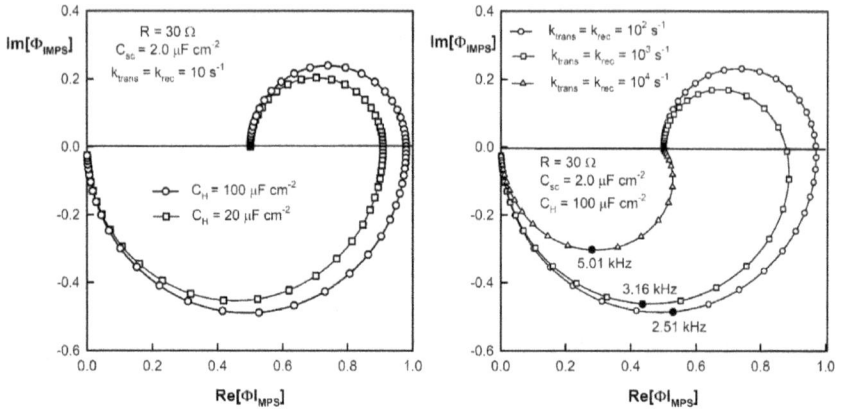

Figure 10.4.   Left: Nyquist plots showing the effects of high-frequency attenuation on the IMPS response. The semicircle in the lower complex plane can be used to find the space charge capacitance if the series resistance is known, for example, from impedance measurements. Note that the IMPS plot crosses the real axis at a value that is less than 1. The effect becomes more pronounced as the values of C_{sc} and C_H become closer. Right: As $(k_{trans} + k_{rec})$ approaches $1/RC$, the upper and lower semicircles begin to merge so that reliable determination of the kinetics is no longer possible.

In the limit that $C_{sc} \ll C_H$, equation (10.31a) becomes

$$\Phi_{IMPS} = \frac{k_{trans} + j\omega}{k_{trans} + k_{rec} + j\omega} \left(\frac{1}{1 + j\omega RC} \right) \tag{10.31c}$$

The RC attenuation of the IMPS signal becomes a serious problem if the series resistance is high, for example, due to poor ohmic contacts to the semiconductor, or if the semiconductor is so highly doped that its space charge capacitance becomes comparable with the Helmholtz capacitance. Figure 10.4 illustrates some of the effects of RC attenuation on the IMPS signature.

10.6   Using IMPS to Study Photoelectrochemical Hydrogen Evolution of p-Si

The simplest IMPS behaviour is seen for carefully prepared single-crystal semiconductor photoelectrodes. A good example is low-doped-type silicon in slightly acidic ammonium fluoride solutions. The surface of the silicon produced by fluoride etching of the surface oxide is hydrogen terminated and almost free of surface states. Under illumination, the p-type silicon behaves as a photocathode. Figure 10.5 illustrates IMPS responses measured at two potentials in the photocurrent onset region (the inset shows

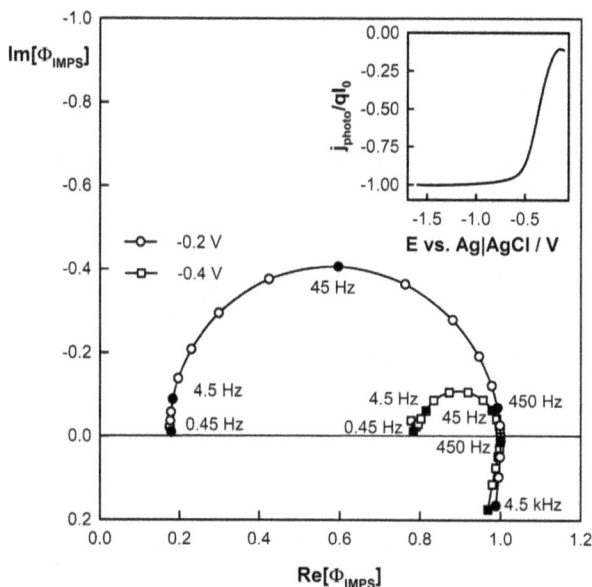

Figure 10.5. IMPS plots for hydrogen evolution on a p-Si photocathode. The inset shows the photocurrent voltage plot corresponding to the intensity of the dc component used in the IMPS measurements. The low value of $\omega_{max} = (k_{trans} + k_{rec})$ indicates that the kinetics of electron transfer and recombination are slow. Reliable values of k_{trans} and k_{rec} can be obtained using the low-frequency intercepts of the IMPS plots and the corresponding values of ω_{max} because RC attenuation is negligible in this instance as the silicon is low doped (i.e., C_{sc} is small).

the steady-state photocurrent voltage plot). The IMPS plots have been normalized by dividing by the modulated Gärtner electron current to obtain the IMPS functions for direct comparison with theory. The RC semicircle cannot be seen in full because the space charge capacitance of the p-doped silicon electrode is very low (of the order of nanofarads). The upper semicircle is observed at much lower frequencies, indicating that the photoelectrochemical hydrogen evolution reaction is slow. This process has also been studied in detail by light-modulated microwave reflectance spectroscopy (LMMR), and further details can be found in Cass *et al.* (2003).

In this particular study, analysis of the IMPS and LMMR responses gave values of k_{trans} and k_{rec} as functions of potential and light intensity. The results have been related to possible mechanisms involving intermediate species, but to achieve an in-depth understanding, further work is needed to relate the results' dynamic measurements such as IMPS and PEIS to time or frequency-resolved spectroscopic measurements. For example, time-resolved Fourier-transform infrared (FTIR) spectroscopy should be able to follow the dynamics of surface-bound hydrogen species on illuminated p-Si and on compound semiconductors such as p-GaAs and p-InP. Perhaps in the future, frequency-resolved light-modulated FTIR will be developed to allow direct comparison with IMPS, PEIS, LMMR and other light-modulated methods.

10.7 Using IMPS to Study Photoelectrochemical Oxygen Evolution on Haematite

Unfortunately, well-behaved single-crystal semiconductors like n-Si cannot be used as photoanodes for photoelectrochemical oxygen evolution because they are easily oxidized under illumination. Instead, measurements have been made on stable transition metal oxides such as TiO_2 and WO_3. However, these wide band gap materials are of little practical interest since they absorb only in the ultraviolet (UV) region. The most widely studied oxide with suitable light absorption is haematite (α-Fe_2O_3), which is an earth abundant n-type metal oxide. It has a band gap of around 2 eV, which is suitable for light-driven water splitting. Unfortunately, its bulk and surface properties are far from ideal, and without modification it is a remarkably poor photoanode. The very short lifetime of photogenerated holes combined with low hole mobility and high n-type doping are severe drawbacks that are difficult to overcome. Most of the research on haematite photoelectrodes has focussed on improving their solid-state and surface properties with the objective of achieving useful solar efficiencies. Two approaches have been moderately successful. The first is to nanostructure the electrode. The rationale behind this is that nanostructuring makes it easier for holes to reach the surface. The second is to modify the surface of the haematite by a so-called co-catalyst such as a cobalt oxide. Even with these improvements, the performance of haematite photoanodes is too low to be of practical interest. Nevertheless, studies of haematite electrodes have enabled us to look deeper into the mechanisms and kinetics of light-driven water splitting reactions. In this section, we will look at some results obtained in our laboratory as part of a collaboration with the group of Professor Upul Wijayantha at Loughborough University. Further details can be found in Peter *et al.* (2012).

For this study, smooth polycrystalline haematite films were grown on fluorine-doped tin oxide (FTO)-coated glass by aerosol-assisted chemical vapour deposition using ferrocene as a precursor. IMPS measurements

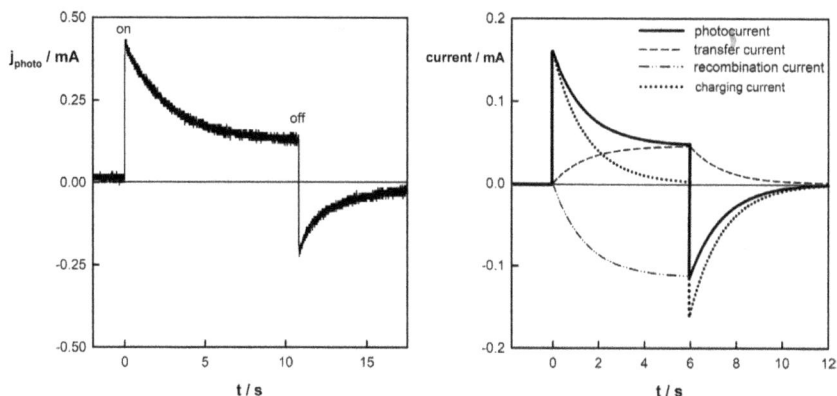

Figure 10.6. Left: A typical photocurrent transient measured for POER on a haematite electrode in alkaline solution (1.0 M NaOH). The decay and overshoot indicate that recombination is occurring, most likely via the Fe=O intermediate in the oxygen evolution reaction. Right: Calculated transient showing the deconvolution into charging, recombination and charge transfer currents (see equations (10.20) and (10.21)).

were made under potentiostatic control in a 3-electrode cell using a single LED (455 nm) controlled by an LED driver. The electrolyte was 1.0 M NaOH. Only a small selection of results for untreated haematite electrodes is presented here to illustrate the application of IMPS.

We can get an idea of how slow the kinetics of transfer and recombination for the POER on haematite are by looking at the photocurrent response to an on/off illumination sequence. Figure 10.6 is a typical photocurrent transient. The almost instantaneous rise in current when the light is switched on is due to the flow of holes towards the interface, i.e., it is a charging current. The risetime is determined by the RC time constant, so it corresponds to the RC semicircle in the IMPS response. The subsequent much slower photocurrent decay can be interpreted in two ways. The simplest is that the decay is due to an increasing electron current due to recombination with holes that are 'queuing' at the surface to react with water. This corresponds to the simple model represented by Figure 10.5. A more realistic interpretation is that the electron flow is due to electrons reacting with the Fe=O that is built up in the first step in the four-hole oxidation of water (we can think of Fe=O as a 'surface-trapped hole'). This process also results in electron–hole-recombination. Eventually, the photocurrent reaches a plateau when a steady state has been reached in which the surface concentration of holes (or Fe=O) is constant. When the illumination is interrupted, the current switches sign to negative before decaying to zero. This negative current is the electron flux

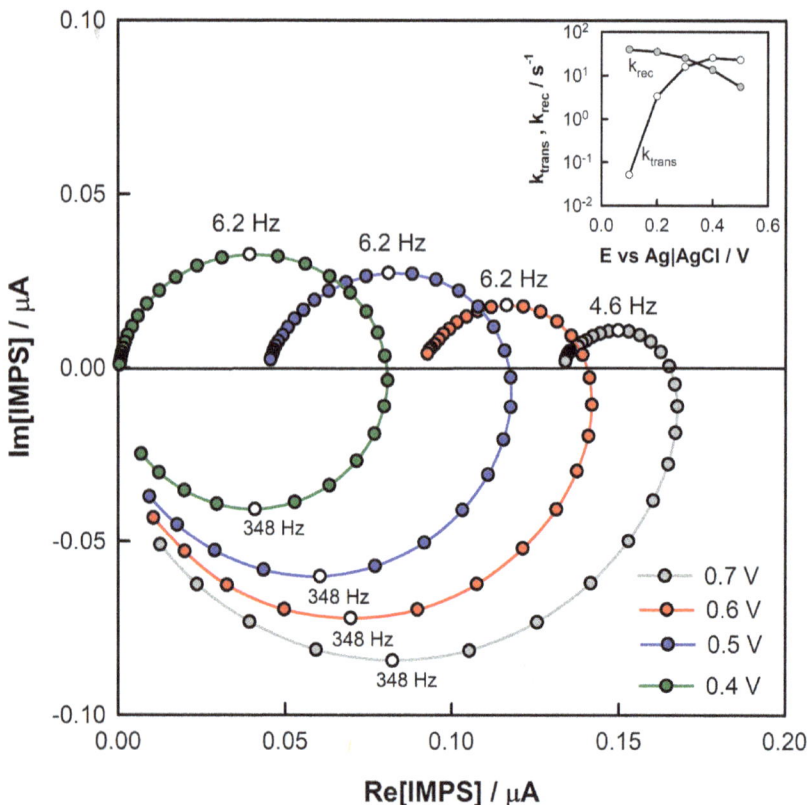

Figure 10.7. IMPS plots for a haematite electrode in 1.0 M NaOH at the potentials shown. The inset shows the values of k_{trans} and k_{rec} derived from the IMPS responses. The RC semicircles correspond to a potential-independent capacitance of 15 μF, indicating that the electrode is behaving non-ideally due to Fermi level pinning. This is confirmed by the strong dependence of k_{trans} on electrode potential. The shift in the crossing point between the upper and lower semicircles is mainly due to the potential dependence of the hole flux j_G predicted by the Gärtner equation when the penetration depth of the illumination is much higher than the width of the space charge region, although RC attenuation may also be important.

into the surface as 'recombination' proceeds, either directly as a electron–hole recombination or, as is more likely, via reduction of the Fe=O built up during the illumination phase. Figure 10.6 also shows the deconvolution of a calculated transient into contributions from charging, charge transfer and recombination. Further details can be found in Peter *et al.* (2020).

Figure 10.7 shows a set of IMPS measurements made on a haematite electrode at different potentials. The general shapes resemble the plots calculated considering the effect of the RC attenuation of the IMPS signal (Figure 10.4), although the positive and negative arcs are flattened to a significant extent. This highlights the problems associated with using heterogeneous polycrystalline photoelectrodes rather than well-defined single crystal semiconductor surfaces. Fitting of the flattened arcs can be achieved by considering a distribution of time constants, but this is a reasonably satisfactory IMPS analysis that can be based on the high- and low-frequency intercepts together with ω_{max} for the upper semicircle. In this case of the IMPS responses shown in Figure 10.7, the 'fit semicircle' routine in ZView® was used to obtain the required values. The derived values of k_{trans} and k_{rec} are shown in the inset to the figure.

The RC semicircle was also analyzed using the fit semicircle function in ZView®. The series resistance determined by PEIS was 28 Ω. And this gives a high value of 15 μF for the capacitance $C = C_{sc}C_H/(C_{sc} + C_H)$, indicating that the haematite is heavily doped so that the space charge capacitance is orders of magnitude higher than in the previous example of p-Si. Remarkably, the C value is independent of electrode potential over the potential range of the IMPS measurement. This behaviour indicates a phenomenon referred to as *Fermi level pinning*, which means that the voltage drop in the semiconductor stays almost constant while the voltage drop across the Helmholtz layer increases as the electrode potential is made more positive. The effect arises from the storage of charge on the surface, and this is normally attributed to the presence of surface states – surface electronic energy levels located in the band gap arising from surface defects. In the present case, it is likely that the surface charge is a consequence of the build-up of surface-bound intermediates such as Fe=O generated by illumination. Under these conditions, the photoelectrode behaves almost like a metal, and the charge transfer rate constant increases as the electrode potential is made more positive.

The rate constants obtained by analysis of the IMPS response also show that the haematite electrode is not behaving as a classical semiconductor electrode. The steep increase in k_{trans} with potential is what we expect for a metal electrode. It indicates that the hole transfer reaction is accelerated by increasing the potential drop across the Helmholtz layer. By contrast, k_{rec} does not decrease as steeply as expected for an ideal photoelectrode without surface states. Both observations are consistent with Fermi level pinning. After treatment of the haematite electrode with a dilute solution of cobalt sulphate, the haematite electrodes behave much more ideally.

10.8 PEIS Response of Photoelectrodes

The approach used to derive the IMPS response of photoelectrodes can be extended to PEIS. We begin by considering the effect of potential modulation of an illuminated photoelectrode. First, the modulation will perturb the width of the space charge region, W_{sc} (equation (10.8)), and hence the flux of holes to the surface given by the Gärtner equation (equation (10.10)). Second, the modulation will perturb the electron Fermi level at the surface and hence the recombination rate constant, k_{rec}. We will assume that the rate constant for charge transfer is not perturbed provided that the majority of the potential drop occurs in the space charge layer of the semiconductor electrode. This assumption will not be valid if the surface charge on the electrode associated with surface states is significant. The derivation of the PEIS response involves considering the resulting perturbation of the charge on the space charge and Helmholtz capacitances. A full derivation is given in Appendix 10.1 at the end of this chapter. The derivation involves relating the time-dependent charges on the space charge and Helmholtz capacitances to the modulated current in the external circuit. The total modulated voltage drop across the electrode (excluding the series resistance) is then obtained using the fact that for each capacitor: (a) the modulated voltage is equal to the modulated charge divided by the differential capacitance; (b) the impedance is the modulated voltage divided by the modulated current.

$$\delta V(t) = \frac{\delta Q(t)}{C} \quad \text{and} \quad Z = \frac{\delta V(t)}{\delta j(t)} \tag{10.32}$$

Our small amplitude equivalent circuit is therefore a series connection of two capacitances and a resistance as shown in Figure 10.8.

At this point, it is important to distinguish between the different types of charges stored in the space charge and Helmholtz capacitances. We will consider an n-type semiconductor electrode in the dark and start with the bulk. Here the space charge region formed in the dark by applying a potential more positive than the flat band potential consists of immobile positively charged donor atoms that have lost an electron. The lower the doping density, the wider the space charge region. The positive charge in the semiconductor is balanced by an equal negative ionic charge in the electrolyte. Now consider the surface of the electrode. If the electrode is an oxide or has an oxide layer on its surface, the acid/base equilibria involving surface –OH groups give rise to an ionic surface charge will depend on the pH. Again the counter charge is located in the solution. Now consider

$$\delta V_{sc}(t) = \frac{\delta Q_{sc}(t)}{C_{sc}} \qquad \delta V_H(t) = \frac{\delta Q_H(t)}{C_H}$$

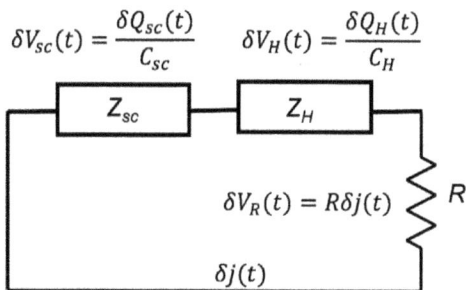

$$\delta V_R(t) = R\delta j(t)$$

Figure 10.8. Small amplitude equivalent circuit for PEIS. The impedances associated with the space charge and Helmholtz layers are derived by considering the modulation of (a) the recombination rate constant, (b) the width of the space charge region and hence the hole flux. In the subsequent discussion of the PEIS impedance, the series resistance is omitted.

the same electrode under illumination and held at a given potential by a potentiostat. Illumination creates electron–hole pairs that are separated, with electrons moving to the back contact and holes, towards the surface. Electron flow in the external circuit will charge the solution side of the Helmholtz capacitance negatively. At any time, electroneutrality demands that the positive hole charge at or near the surface should be balanced by the sum of the electron charge on the other side of the space charge layer and the net additional negative charge on the Helmholtz layer. The two situations – dark and illuminated – are illustrated schematically in Figure 10.9. The figure shows that current flow in the external circuit decreases the charge on the space charge capacitance and increases the charge on the Helmholtz capacitance.

So far, we have not considered what happens to holes at the surface. If a hole accepts an electron from a solution species, the charge on the Helmholtz capacitance will decrease. For example, if we imagine that a single hole at the interface is balanced by a single $Fe(CN)_6^{4-}$ ion in the Helmholtz layer, after electron transfer the hole will have disappeared and the 4- ionic charge will have been replaced by the 3-charge on the oxidation product, $Fe(CN)_6^{3-}$. In summary, we can relate changes in the charge on the space charge and Helmholtz capacitances to (a) current flow in the external circuit and (b) to hole transfer across the interface. This is the starting point for the derivation of the PEIS expressions in Appendix 10.1.

The total hole excess charge is balanced by the *sum* of the electron charge on the left-hand side of the space charge capacitance and the excess negative ionic charge on the right-hand side of the Helmholtz capacitance.

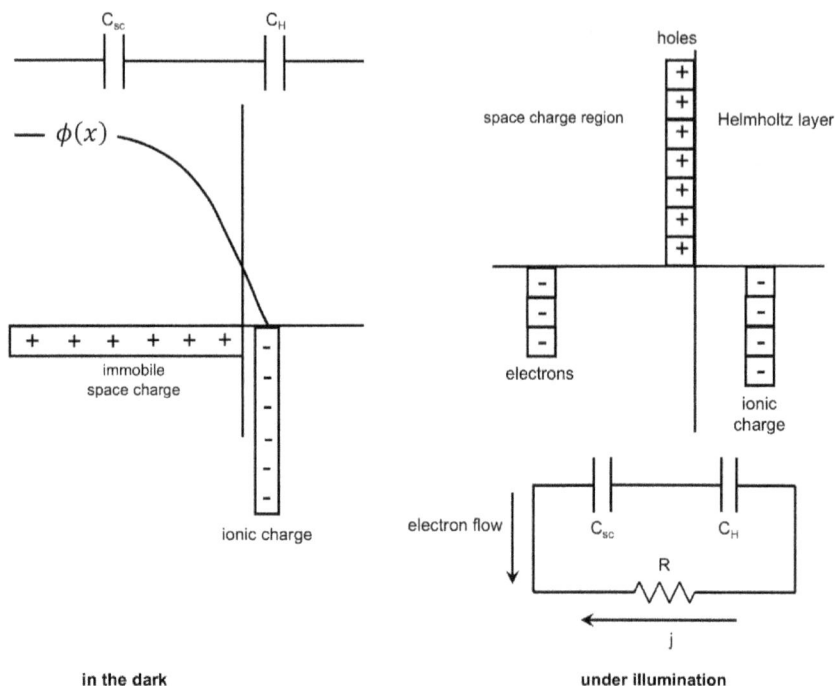

Figure 10.9. Left: Potential distribution across the interface between an n-type semiconductor and the electrolyte under conditions where a space charge layer is formed by applying a potential more positive than the flat band potential. For simplicity, it has been assumed that solution pH is chosen so that the net surface ionic charge is zero. This pH corresponds to the isoelectric point. Right: Illumination creates equal numbers of electrons and holes. Current flow in the external circuit corresponds to electrons leaving the semiconductor and passing via the counter electrode to create an additional negative ionic charge on the right-hand side of the Helmholtz layer.

Hole transfer across the interface (not shown) decreases the charge on the Helmholtz layer.

Appendix 10.1 shows that for most applications, we can neglect the effects of modulating the width of the space charge region. The contribution of the Helmholtz impedance to the total impedance is also small enough to ignore, even in the case of highly doped semiconductors. This means that the PEIS impedance can be adequately represented by the expression for the space charge impedance

$$Z_{\text{PEIS}} = \frac{1}{j\omega C_{\text{sc}} + \frac{qj_G}{k_BT} \cdot \frac{k_{\text{rec}}}{k_{\text{trans}}+k_{\text{rec}}} \cdot \frac{k_{\text{trans}}+j\omega}{k_{\text{trans}}+k_{\text{rec}}+j\omega}} \tag{10.33a}$$

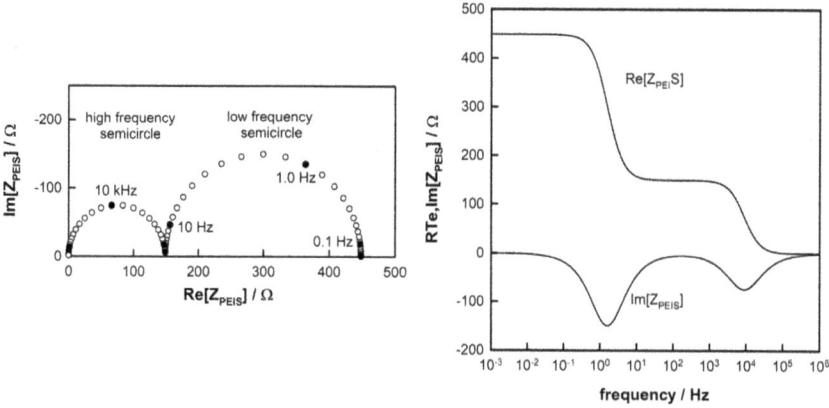

Figure 10.10. Nyquist and Bode plots of the PEIS impedance calculated from equation (10.33a). Parameter values: $N_d = 10^{17}\,\text{cm}^{-3}$, $\varepsilon = 10$, $C_H = 20\,\mu\text{F}\,\text{cm}^{-2}$, $\alpha = 10^4\,\text{cm}^{-1}$, $L_p = 10^{-5}\,\text{cm}$, $I_0 = 10^{16}\,\text{cm}^{-2}\,\text{s}^{-1}$, $E - E_{\text{fb}} = 0.5\,\text{V}$, $k_{\text{trans}} = 10\,\text{s}^{-1}$, $k_{\text{rec}} = 20\,\text{s}^{-1}$. Calculated values: $W_{\text{sc}} = 7.4 \times 10^{-7}\,\text{cm}$, $C_{\text{sc}} = 1.19 \times 10^{-7}\,\text{F}\,\text{cm}^{-2}$, $j_G = 0.25\,\text{mA}\,\text{cm}^{-2}$, EQE $= 0.156$.

This expression describes *two* semicircles in the Nyquist plot. We can obtain some information about these semicircles by considering limiting forms. The first point to note is that in the dark, the impedance given by equation (10.33a) reduces to $-j/\omega C_{\text{sc}}$, i.e., just the impedance of the space charge capacitance. Next, to obtain the low-frequency intercept under illumination, consider the limit $\omega \to 0$, when equation (10.32) reduces to

$$Z_{\text{PEIS},\omega \to 0} = \frac{(k_{\text{trans}} + k_{\text{rec}})^2}{\frac{qj_G}{k_BT}k_{\text{trans}}k_{\text{rec}}} \qquad (10.33b)$$

At mid frequencies, when $\omega \gg k_{\text{trans}}, k_{\text{rec}}$, but $j\omega C_{\text{sc}} \ll \frac{qj_G}{k_BT} \cdot \frac{k_{\text{rec}}}{k_{\text{trans}}+k_{\text{rec}}}$, the impedance tends towards the real value.

$$\text{Re}(Z_{\text{PEIS,mf}}) = \frac{1}{\frac{qj_G}{k_BT} \cdot \frac{k_{\text{rec}}}{k_{\text{trans}}+k_{\text{rec}}}} \qquad (10.33c)$$

Finally, at very high frequencies, the impedance tends to zero, which means that the measured PEIS impedance will just be the series resistance, R.

Figure 10.10 shows a Nyquist plot of the PEIS impedance calculated for some typical values of the variables in equation (10.33a). It shows the two semicircles with low and medium frequency values of the real component. It is easy to show that the ratio of the low and medium frequency real components is given by

$$\frac{Z_{\text{PEIS},\omega\to 0}}{\text{Re}(Z_{\text{PEIS,mf}})} = 1 + \frac{k_{\text{rec}}}{k_{\text{trans}}} \qquad (10.34)$$

The radial frequency of the maximum in the low-frequency semicircle is equal to the transfer rate constant ($\omega_{\text{max,LF}} = k_{\text{trans}}$), and this allows us to determine k_{rec} from equation (10.34).

Finally, the radial frequency of the maximum in the high-frequency semicircle is given by

$$\omega_{\text{max,HF}} = \frac{q j_G}{k_B T C_{\text{sc}}} \frac{k_{\text{rec}}}{k_{\text{trans}} + k_{\text{rec}}} \qquad (10.35)$$

which allows us to find the space charge capacitance if j_G is known.

The treatment of IMPS and PEIS given in this chapter is based on a kinetic model. The original theory was validated by IMPS and PEIS measurements of photoelectrochemical hydrogen evolution on p-InP, which demonstrated that the analysis of the results obtained using the two methods gave identical values of k_{trans} and k_{rec}. Further details can be found in Ponomarev and Peter (1995). Alternative approaches based on RC equivalent circuits are frequently used to model PEIS data in the literature. For example, the two semicircles shown in Figure 10.10 have been described by a series connection of two parallel RC circuits, with the capacitances attributed to the space charge and Helmholtz capacitances and the parallel resistances attributed to recombination and charge transfer or to surface states. These interpretations are not easy to reconcile with the consideration of the physical processes that are analyzed in Appendix 10.1, so they are not pursued here.

In this final part of Chapter 10, we look at some results obtained in collaboration with the group of Professor Upul Wijayantha (Loughborough University) for the same haematite electrodes that were used for the IMPS measurements described in Section 10.7. Figure 10.11 shows an example of the PEIS response and the fitting to equation (10.33a). The two semicircles can be distinguished, but they are flattened and merged. This effect was also seen in the IMPS response of the same electrodes (see Figure 10.7). In spite of this, analysis of the data again gave results that are compatible with those derived from IMPS. Further details can be found in Wijayantha *et al.* (2011).

We will finish this section with a health warning. IMPS and PEIS have been widely used to study the complex multistep reactions that are involved in water splitting. The analysis of the data in terms of the

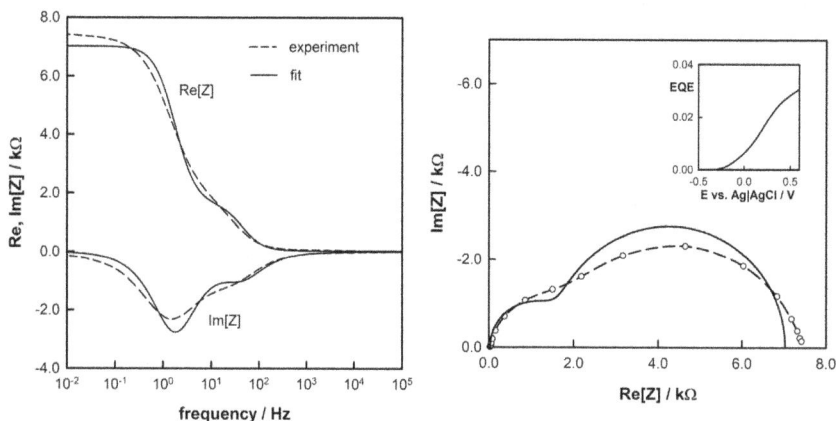

Figure 10.11. Bode and Nyquist plots illustrating the PEIS response of a haematite electrode in 1.0 M NaOH. Illumination 455 nm, $11 \, \mathrm{mW \, cm^{-2}}$. $E = 0 \, \mathrm{V}$ vs. Ag | AgCl. The inset shows the steady-state potential dependence of the external quantum efficiency (EQE) under these conditions.

phenomenological model is usually straightforward, and some indication of possible mechanisms can be obtained, for example, from the intensity dependence of k_{trans} and k_{rec}. However, any proposed mechanistic scheme needs to be validated by other methods, for example, infrared spectroscopy.

Appendix 10.1 Derivation of PEIS Expressions

The objective of this appendix is to illustrate how the PEIS impedance is derived by considering the modulation of the charge on the space charge and Helmholtz capacitances arising from charging, charge transfer and recombination. The derivation looks rather lengthy, but that is because I have tried to show each step as clearly as possible.

We begin by considering how potential modulation affects the different contributions to the net current in the external circuit. Potential modulation changes the electron Fermi level and hence the recombination current due to electrons flowing from the bulk of the semiconductor to the surface region where they combine with holes. At the same time, potential modulation alters the width of the space charge region, and hence the hole current predicted by the Gärtner equation. This effect will be important if light penetrates well beyond the edge of the SCR. In summary, we need to consider the modulation of both the electron current *and* the hole current as well as the current in the external circuit.

Consider a small modulation of W_{sc} by modifying the expression in the main text.

$$W_{sc} + \delta W_{sc} = \left(\frac{2(V_{sc} + \delta V_{sc})\varepsilon\varepsilon_0}{qN_d}\right)^{1/2} = \left(\frac{2V_{sc}\varepsilon\varepsilon_0}{qN_d}\right)^{1/2}\left(1 + \frac{\delta V_{SC}}{V_{sc}}\right)^{1/2} \tag{A10.1}$$

Here, δV_{sc} is the sinusoidal modulation of the voltage across the space charge region. Of course, this is *not* the same as the total voltage modulation across the semiconductor and series resistance. Expanding the second term for a small perturbation

$$W_{sc} + \delta W_{sc} = W_{sc}\left(1 + \frac{\delta V_{sc}}{2V_{sc}}\right) \tag{A10.2}$$

The modulated component we want is therefore

$$\delta W_{sc} = W_{sc}^0 \frac{\delta V_{sc}}{2V_{sc}} \tag{A10.3}$$

where W_{sc}^0 is the unperturbed width of the SCR.

Now consider the effect of the modulation of W_{sc} on the hole current predicted by the Gärtner equation. Again, we use an expansion for small arguments – in this case, the argument of the exponential term

$$j_G + \delta j_G = qI_0\left(1 - \frac{e^{-\alpha(W_{sc}^0 + \delta W_{sc})}}{1 + \alpha L_p}\right) = qI_0\left(1 - \frac{e^{-\alpha W_{sc}^0}}{1 + \alpha L_p}(1 - \alpha\delta W_{sc})\right) \tag{A10.4}$$

from which it follows that the ac component of the hole current is

$$\delta j_G = qI_0\frac{e^{-\alpha W_{sc}^0}}{1 + \alpha L_p}\alpha\delta W_{sc} = (qI_0 - j_G)\alpha\delta W_{sc} \tag{A10.5}$$

Substituting for δW_{sc} from equation (A10.3) and noting that $\delta V_{sc} = \delta Q_{sc}/C_{sc}$ gives

$$\delta j_G = \frac{\delta V_{sc}}{2V_{sc}} \cdot \alpha W_{sc}^0(qI_0 - j_G) = \frac{\delta Q_{sc}}{2V_{sc}C_{sc}}\alpha W_{sc}^0(qI_0 - j_G)$$

$$= \frac{\delta Q_{sc}(C_H + C_{sc})}{2VC_HC_{sc}}\frac{\alpha\varepsilon\varepsilon_0}{C_{sc}}(qI_0 - j_G) \tag{A10.6}$$

We can simplify this equation by writing it in the form

$$\delta j_G = \Gamma\delta Q_{sc} \text{ with } \Gamma = \frac{\alpha\varepsilon\varepsilon_0(C_H + C_{sc})(qI_0 - j_G)}{2VC_HC_{sc}^2}(\text{units s}^{-1}) \tag{A10.7}$$

Note the term $(qI_0 - j_G)$. If the penetration depth of the illumination is much smaller than the width of the space charge region, the Gärtner equation shows us that j_G will be close to qI_0, so Γ will be small so that we can forget about the effect on j_G of modulating the width of the SCR.

Now consider how modulation of V_{sc} affects the recombination rate constant k_{rec}. Without modulation, the recombination rate constant k_r^0 depends on the concentration of electrons at the surface, which in turn falls exponentially as the band bending is increased.

$$k_r^0 = k_{rec}' n_{x=0} = k_{rec}' N_d e^{-\frac{\beta q V_{sc}}{k_B T}} \tag{A10.8}$$

Here, k_{rec}' is a second-order rate constant, N_d is the doping density and β is a non-ideality factor. For a small modulation δV_{sc},

$$k_{rec}^0 + \delta k_{rec} = k_{rec}' N_d e^{-\frac{\beta q (V_{sc} + \delta V_{sc})}{k_B T}} \tag{A10.9}$$

From which we obtain using the expansion $e^{-x} \cong 1 - x$ for $x \ll 1$

$$\delta k_{rec} = -k_{rec}^0 \frac{\beta q \delta V_{sc}}{k_B T} \tag{A10.10}$$

Now consider the perturbation of charge on the space charge capacitance. For small perturbation of voltage, we can consider the space charge capacitance to be constant, so that we can replace δV_{sc} by $\delta Q_{sc}/C_{sc}$ to obtain

$$\delta k_r = -k_r^0 \frac{q}{k_B T} \frac{\delta Q_{sc}}{C_{sc}} \tag{A10.11}$$

Now consider the perturbation of the rate of change in the surface concentration of holes, which depends on j_G, the hole current into the surface and the rates of removal of holes by charge transfer and recombination. If p^0 is the steady-state hole concentration under illumination and δp is the small perturbation in the PEIS experiment, then

$$\frac{d(p^0 + \delta p)}{dt} = j_G + \delta j_G - k_{trans}(p^0 + \delta p) - (k_{rec}^0 + \delta k_{rec})(p^0 + \delta p) \tag{A10.12}$$

Here, k_{rec}^0 is the dc (i.e., unperturbed) value of the recombination rate constant.

Multiplying out the brackets and neglecting the small cross-term $\delta k_r \delta p$ gives

$$\frac{d(p^0 + \delta p)}{dt} = j_G + \delta j_G - k_{\text{trans}} p^0 - k_{\text{rec}}^0 p^0 - k_{\text{rec}}^0 \delta p - \delta k_{\text{rec}} p^0 \quad \text{(A10.13)}$$

It follows that

$$\frac{d\delta p}{dt} = \delta j_G - (k_{\text{trans}} + k_{\text{rec}}^0)\delta p - \delta k_{\text{rec}} p^0 \quad \text{(A10.14)}$$

Note that δp is a time-dependent quantity – I have left out the (t) to save space. Our three time-dependent variables are all sinusoidal, but phase-shifted relative to each other, so they can be expressed as

$$\delta p(t) = |\delta p| e^{j\omega t} \quad \text{(A10.15a)}$$

$$\delta j_G = |\delta j_G| e^{j\omega t + \phi} \quad \text{(A10.15b)}$$

$$\delta k_{\text{rec}} = |\delta k_{\text{rec}}| e^{j\omega t + \theta} \quad \text{(A10.15c)}$$

From equation (A10.12), noting that $\frac{d}{dt} e^{j\omega t + \phi} = j\omega e^{j\omega t + \phi}$, we obtain

$$j\omega \delta p = \delta j_G - (k_{\text{trans}} + k_{\text{rec}}^0)\delta p - \delta k_{\text{rec}} p^0 \quad \text{(A10.16)}$$

Collecting up terms in δp gives

$$\delta p(k_{\text{trans}} + k_{\text{rec}}^0 + j\omega) = \delta j_G + k_{\text{rec}}^0 \frac{q}{k_B T} \frac{\delta Q_{\text{sc}}}{C_{\text{sc}}} p^0$$

$$= \delta j_G + \frac{q j_G \delta Q_{\text{sc}}}{k_B T C_{\text{sc}}} \cdot \frac{k_{\text{rec}}^0}{k_{\text{trans}} + k_{\text{rec}}^0} \quad \text{(A10.17)}$$

which allows us to obtain the modulated component of the surface hole concentration.

$$\delta p = \frac{\boldsymbol{\delta j_G}}{\boldsymbol{k_{\text{trans}} + k_{\text{rec}}^0 + j\omega}} + \frac{q j_G \delta Q_{\text{sc}}}{k_B T C_{\text{sc}}} \frac{k_{\text{rec}}^0}{(k_{\text{trans}} + k_{\text{rec}}^0)} \cdot \frac{1}{(k_{\text{trans}} + k_{\text{rec}}^0 + j\omega)}$$

$$\text{(A10.18)}$$

Note the additional first term (in **bold**) arising from modulation of the Gartner current.

Now consider the change in the charge in the SCR. Since the separation of photogenerated electrons and holes in the SCR illumination gives rise to a photovoltage, the band bending is reduced, and the space charge decreases. Q_{sc} is therefore decreased by the hole current and increased by the recombination current and the current flowing in the external circuit, which corresponds to electrons leaving via the back contact. We can express this in terms of the modulated components as

$$-\frac{d\delta Q_{sc}}{dt} = -j\omega \delta Q_{sc} = \delta j_G - k_{rec}^0 \delta p - \delta k_{rec} p^0 - \delta j \qquad (A10.19)$$

Again, note the δj_G term arising from modulation of the width of the SCR.

Now we are ready to substitute the expressions for δp and δk_r and rearrange the terms.

$$j\omega \delta Q_{sc} = -\delta j_G + k_r^0 \left(\frac{\delta j_G}{k_{trans} + k_{rec}^0 + j\omega} \right.$$

$$+ \frac{qj_G \delta Q_{sc}}{k_B T C_{sc}} \cdot \frac{k_{rec}^0}{k_{trans} + k_{rec}^0} \cdot \frac{1}{k_{trans} + k_{rec}^0 + j\omega} \Bigg)$$

$$- \frac{qj_G \delta Q_{sc}}{k_B T C_{sc}} \cdot \frac{k_{rec}^0}{(k_{trans} + k_{rec}^0)} + \delta j \qquad (A10.20)$$

$$j\omega \delta Q_{sc} = -\delta j_G \left(1 - \frac{k_{rec}^0}{k_{trans} + k_{rec}^0 + j\omega} \right)$$

$$+ \frac{qj_G \delta Q_{sc}}{k_B T C_{sc}} \cdot \frac{k_{rec}^0}{k_{trans} + k_{rec}^0} \cdot \frac{k_{rec}^0}{k_{trans} + k_{rec}^0 + j\omega}$$

$$- \frac{qj_G \delta Q_{sc}}{k_B T C_{sc}} \cdot \frac{k_{rec}^0}{(k_{trans} + k_{rec}^0)} + \delta j \qquad (A10.21)$$

$$j\omega \delta Q_{sc} = -\delta j_G \left(1 - \frac{k_{rec}^0}{k_{trans} + k_{rec}^0 + j\omega} \right) - \frac{qj_G \delta Q_{sc}}{k_B T C_{sc}} \cdot \frac{k_{rec}^0}{k_{trans} + k_{rec}^0}$$

$$\times \left(1 - \frac{k_r^0}{k_{trans} + k_{rec}^0 + j\omega} \right) + \delta j \qquad (A10.22)$$

$$j\omega \delta Q_{sc} = -\Gamma \delta Q_{sc} \frac{k_{trans} + j\omega}{k_{trans} + k_{rec}^0 + j\omega}$$

$$- \frac{qj_G \delta Q_{sc}}{k_B T C_{sc}} \cdot \frac{k_{rec}^0}{k_{trans} + k_{rec}^0} \cdot \frac{k_t + j\omega}{k_{trans} + k_{rec}^0 + j\omega} + \delta j \qquad (A10.23)$$

$$\delta Q_{sc} \left(i\omega + \Gamma \frac{k_{trans} + j\omega}{k_{trans} + k_{rec}^0 + j\omega} \right.$$

$$+ \frac{qj_G}{k_B T C_{sc}} \frac{k_{rec}^0}{k_{trans} + k_{rec}^0} \cdot \frac{k_{trans} + j\omega}{k_{trans} + k_{rec}^0 + j\omega} \right) = \delta j \qquad (A10.24)$$

$$\delta Q_{\rm sc} = \frac{\delta j}{\left(j\omega + \Gamma \frac{k_{\rm trans}+j\omega}{k_{\rm trans}+k_{\rm rec}^0+j\omega} + \frac{qj_G}{k_BTC_{\rm sc}} \cdot \frac{k_{\rm rec}^0}{k_{\rm trans}+k_{\rm rec}^0} \cdot \frac{k_{\rm trans}+j\omega}{k_{\rm trans}+k_{\rm rec}^0+j\omega}\right)}$$

$$(A10.25)$$

This is exactly what we want. We can define the impedance of the space charge region as

$$Z_{\rm sc} = \frac{\delta V_{\rm sc}}{\delta j} = \frac{1}{C_{\rm sc}} \frac{\delta Q_{\rm sc}}{\delta j}$$

$$= \frac{1}{C_{\rm sc}\left(j\omega + \Gamma \frac{k_{\rm trans}+j\omega}{k_{\rm trans}+k_{\rm rec}^0+j\omega} + \frac{qj_G}{k_BTC_{\rm sc}} \cdot \frac{k_{\rm rec}^0}{k_{\rm trans}+k_{\rm rec}^0} \cdot \frac{k_{\rm trans}+j\omega}{k_{\rm trans}+k_{\rm rec}^0+j\omega}\right)}$$

$$(A10.26a)$$

giving our final expression for the space charge impedance

$$Z_{\rm sc} = \frac{1}{j\omega C_{\rm sc} + \Gamma C_{\rm sc}\frac{k_{\rm trans}+j\omega}{k_{\rm trans}+k_{\rm rec}^0+j\omega} + \frac{qj_G}{k_BT}\frac{k_{\rm rec}^0}{k_{\rm trans}+k_{\rm rec}^0} \cdot \frac{k_{\rm trans}+j\omega}{k_{\rm trans}+k_{\rm rec}^0+j\omega}}$$

$$(A10.26b)$$

Now we look at $\frac{d\delta Q_H}{dt}$ in order to drive the Helmholtz impedance.

$$\frac{d\delta Q_H}{dt} = j\omega \delta Q_H = -k_{\rm trans}\delta p + \delta j \quad {\rm where} \quad \delta Q_H = |\delta Q_H|e^{j\omega t+\varphi}$$

$$(A10.27)$$

The first term on the right-hand side describes the reduction in the charge on the Helmholtz capacitance when a hole is transferred to the solution. The second term describes the increase in charge on C_H when a conventional positive current flows into the semiconductor via the back contact (i.e., electrons leave via the back contact).

Recalling that

$$\delta p = \frac{\delta j_G}{k_{\rm trans} + k_{\rm rec}^0 + j\omega} + \frac{qj_G\delta Q_{\rm sc}}{k_BTC_{\rm sc}}\frac{k_{\rm rec}^0}{(k_{\rm trans} + k_{\rm rec}^0)} \cdot \frac{1}{(k_{\rm trans} + k_{\rm rec}^0 + j\omega)}$$

$$\delta j_G = \Gamma \delta Q_{\rm sc}$$

$$\delta Q_{\rm sc} = \frac{\delta j}{\left(i\omega + \Gamma \frac{k_t+i\omega}{k_t+k_r^0+i\omega} + \frac{qj_G}{kTC_{\rm sc}} \cdot \frac{k_r^0}{k_t+k_r^0} \cdot \frac{k_t+i\omega}{k_t+k_r^0+i\omega}\right)}$$

we find that

$$j\omega \delta Q_H = -k_{\rm trans}\delta Q_{\rm sc}\left(\frac{\Gamma}{k_{\rm trans} + k_{\rm rec}^0 + j\omega} + \frac{qj_G}{k_BTC_{\rm sc}}\frac{k_{\rm rec}^0}{(k_{\rm trans} + k_{\rm rec}^0)}\right.$$

$$\left. \cdot \frac{1}{(k_{\rm trans} + k_{\rm rec}^0 + j\omega)}\right) + \delta j \qquad (A10.28)$$

$$j\omega\delta Q_H$$

$$= -\frac{\left(\frac{k_{\text{trans}}\Gamma}{k_{\text{trans}}+k_{\text{rec}}^0+j\omega} + \frac{qj_G}{k_BTC_{\text{sc}}}\frac{k_{\text{rec}}^0}{(k_{\text{trans}}+k_{\text{rec}}^0)}\cdot\frac{k_t}{(k_{\text{trans}}+k_{\text{rec}}^0+j\omega)}\right)\delta j}{\left(j\omega + \Gamma\frac{k_{\text{trans}}+j\omega}{k_{\text{trans}}+k_{\text{rec}}^0+j\omega} + \frac{qj_G}{k_BTC_{\text{sc}}}\cdot\frac{k_{\text{rec}}^0}{k_{\text{trans}}+k_{\text{rec}}^0}\cdot\frac{k_{\text{trans}}+j\omega}{k_{\text{trans}}+k_{\text{rec}}^0+j\omega}\right)} + \delta j$$

$$(A10.29)$$

$$j\omega\delta Q_H$$

$$= \left[1 - \frac{\frac{k_{\text{trans}}\Gamma}{k_{\text{trans}}+k_{\text{rec}}^0+j\omega} + \frac{qj_G}{k_BTC_{\text{sc}}}\frac{k_{\text{rec}}^0}{(k_{\text{trans}}+k_{\text{rec}}^0)}\cdot\frac{k_{\text{trans}}}{(k_{\text{trans}}+k_{\text{rec}}^0+j\omega)}}{i\omega + \Gamma\frac{k_{\text{trans}}+j\omega}{k_{\text{trans}}+k_{\text{rec}}^0+j\omega} + \frac{qj_G}{k_BTC_{\text{sc}}}\cdot\frac{k_{\text{rec}}^0}{k_{\text{trans}}+k_{\text{rec}}^0}\cdot\frac{k_{\text{trans}}+j\omega}{k_{\text{trans}}+k_{\text{rec}}^0+j\omega}}\right]\delta j$$

$$(A10.30)$$

$$j\omega\delta Q_H$$

$$= \delta j\cdot\frac{j\omega + \Gamma\frac{i\omega}{k_{\text{trans}}+k_{\text{rec}}^0+j\omega} + \frac{qj_G}{k_BTC_{\text{sc}}}\cdot\frac{k_{\text{rec}}^0}{k_{\text{trans}}+k_{\text{rec}}^0}\cdot\frac{i\omega}{k_{\text{trans}}+k_{\text{rec}}^0+j\omega}}{j\omega + \Gamma\frac{k_t+i\omega}{k_{\text{trans}}+k_{\text{rec}}^0+j\omega} + \frac{qj_G}{k_BTC_{\text{sc}}}\cdot\frac{k_{\text{rec}}^0}{k_{\text{trans}}+k_{\text{rec}}^0}\cdot\frac{k_{\text{trans}}+j\omega}{k_{\text{trans}}+k_{\text{rec}}^0+j\omega}}$$

$$(A10.31)$$

$$\delta Q_H$$

$$= \delta j\cdot\frac{1 + \Gamma\frac{1}{k_{\text{trans}}+k_{\text{rec}}^0+j\omega} + \frac{qj_G}{k_BTC_{\text{sc}}}\cdot\frac{k_{\text{rec}}^0}{k_{\text{trans}}+k_{\text{rec}}^0}\cdot\frac{k_{\text{rec}}^0}{k_{\text{trans}}+k_{\text{rec}}^0}\cdot\frac{1}{k_{\text{trans}}+k_{\text{rec}}^0+j\omega}}{j\omega + \Gamma\frac{k_{\text{trans}}+j\omega}{k_{\text{trans}}+k_{\text{rec}}^0+j\omega} + \frac{qj_G}{k_BTC_{\text{sc}}}\cdot\frac{k_{\text{rec}}^0}{k_{\text{trans}}+k_{\text{rec}}^0}\cdot\frac{k_{\text{trans}}+j\omega}{k_{\text{trans}}+k_{\text{rec}}^0+j\omega}}$$

$$(A10.32)$$

Now we can obtain the Helmholtz impedance.

$$Z_H = \frac{d}{\delta j}\frac{\delta Q_H}{C_H}$$

$$= \frac{1 + \Gamma\frac{1}{k_{\text{trans}}+k_{\text{rec}}^0+j\omega} + \frac{qj_G}{k_BTC_{\text{sc}}}\cdot\frac{k_{\text{rec}}^0}{k_{\text{trans}}+k_{\text{rec}}^0}\cdot\frac{1}{k_{\text{trans}}+k_{\text{rec}}^0+j\omega}}{C_H\left(j\omega + \Gamma\frac{k_{\text{trans}}+i\omega}{k_{\text{trans}}+k_{\text{rec}}^0+j\omega} + \frac{qj_G}{k_BTC_{\text{sc}}}\cdot\frac{k_{\text{rec}}^0}{k_{\text{trans}}+k_{\text{rec}}^0}\cdot\frac{k_{\text{trans}}+i\omega}{k_{\text{trans}}+k_{\text{rec}}^0+j\omega}\right)}$$

$$(A10.33)$$

In the limit that $j_G \to qI_0$, the additional terms containing Γ disappear as expected.

Considering the series connection of two capacitors and the series resistance, we can use Kirchhoff's voltage law

$$\frac{\delta Q_{\text{sc}}}{C_{\text{sc}}} + \frac{\delta Q_H}{C_H} + \delta jR = \delta V \qquad (A10.34)$$

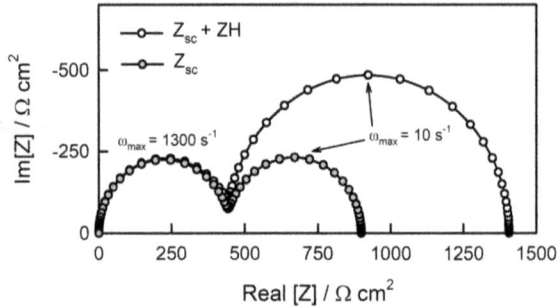

Figure A10.1. Nyquist plots of the space charge impedance Z_{sc}, and total impedance $Z_{sc} + Z_H$ for a highly doped photoelectrode calculated for the following parameter values. $V_{sc} = 0.5\,\mathrm{V}$, $L_p = 10\,\mathrm{nm}$, $N_d = 10^{19}\,\mathrm{cm}^{-3}$, $\varepsilon = 25$, $C_H = 200\,\mu\mathrm{F}\,\mathrm{cm}^{-2}$, $\alpha = 10^5\,\mathrm{cm}^{-1}$, $k_{trans} = k_{rec} = 10\,\mathrm{s}^{-1}$, $I_0 = 10^{16}\,\mathrm{cm}^{-2}\,\mathrm{s}^{-1}$. Note how inclusion of Z_H expands the low frequency semicircle. For such highly doped electrodes, the usual PEIS analysis will give incorrect values of k_{rec}, but k_{trans} values should still be reliable.

$$\frac{\delta V}{\delta j} = R + \frac{\delta Q_{sc}}{\delta j C_{sc}} + \frac{\delta Q_H}{\delta j C_H} = R + Z_{sc} + Z_H \qquad (\text{A10.35})$$

Now we examine the contribution of Z_H to Z_{PEIS}. Figure A10.1 shows that if the semiconductor is so highly doped that its space charge capacitance exceeds $1\,\mu\mathrm{F}\,\mathrm{cm}^{-2}$, the contribution from Z_H becomes significant, expanding the low frequency semicircle without affecting the frequency of its maximum. This can lead to large errors in the determination of the recombination rate constant, but the derived value of the charge transfer rate constant is still reliable.

Now we look at the importance of the terms arising from modulation of the space charge region. To maximize the possible effect, we choose parameters such that the penetration depth of the light is much greater than the width of the space charge region. This means that the term $(j_G - q I_0)$ in the expressions for Z_{sc} and Z_H is large and negative. Figure A10.2 demonstrates that even under these conditions, for most purposes we can neglect the contribution from the modulation of the hole current j_G due to the modulation of W_{sc}.

This rather lengthy discussion shows that the simple analysis of PEIS based on just the charge transfer impedance without the Γ terms is adequate for most purposes. This greatly simplifies the derivation of the rate constants k_{trans} and k_{rec}.

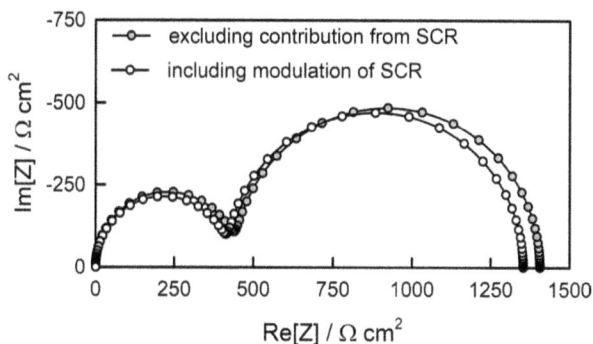

Figure A10.2. Nyquist plots of the total impedance calculated for the parameter values used for Figure A10.1, showing that the effect of including the modulation of the width of the space charge region is quite small. even when the penetration depth of the illumination is considerably greater than the width of the space charge region. The effect can be neglected for routine analysis.

References

Cass, M. J., Duffy, N. W., Peter, L. M., Pennock, S. R., Ushiroda, S. & Walker, A. B. 2003. Microwave reflectance studies of photoelectrochemical kinetics at semiconductor electrodes. 2. Hydrogen evolution at p-Si in ammonium fluoride solution. *Journal of Physical Chemistry B*, 107, 5864–5870.

Gärtner, W. W. 1959. Depletion-layer photoeffects in semiconductors. *Physical Review*, 116, 84–87.

Memming, R. 2015. *Semiconductor Electrochemistry*. Hoboken, NJ: Wiley.

Peter, L. M., Walker, A. B., Bein, T., Hufnagel, A. G. & Kondofersky, I. 2020. Interpretation of photocurrent transients at semiconductor electrodes: Effects of band-edge unpinning. *Journal of Electroanalytical Chemistry*, 872, article number 114234.

Peter, L. M., Wijayantha, K. G. U. & Tahir, A. A. 2012. Kinetics of light-driven oxygen evolution at alpha-Fe_2O_3 electrodes. *Faraday Discussions*, 155, 309–322.

Ponomarev, E. A. & Peter, L. M. 1995. A comparison of intensity-modulated photocurrent spectroscopy and photoelectrochemical impedance spectroscopy in a study of photoelectrochemical hydrogen evolution at p-InP. *Journal of Electroanalytical Chemistry*, 397, 45–52.

Sato, N. 1998. *Electrochemistry at Metal and Semiconductor Electrodes*. Maarssen: Elsevier.

Wijayantha, K. G. U., Saremi-Yarahmadi, S. & Peter, L. M. 2011. Kinetics of oxygen evolution at alpha-Fe_2O_3 photoanodes: A study by photoelectrochemical impedance spectroscopy. *Physical Chemistry Chemical Physics*, 13, 5264–5270.

Answers to Selected Problems

Chapter 1

Problem 1.1

The phase angle $-\pi/4$ corresponds to $-45°$. Remember that the tangent of the phase angle is given by $\tan\theta = \frac{\text{Im}[Z]}{\text{Re}[Z]}$. Since the tangent of $-45°$ is -1, it follows that $\text{Im}[Z] = -\text{Re}[Z]$. We also know that $|z| = 1414$ $\Omega = \sqrt{\text{Re}^2[Z] + \text{Im}^2[Z]}$. Since $\text{Im}[Z] = -\text{Re}[Z]$, we can replace $\text{Re}^2[Z]$ by $\text{Re}^2[Z]$. It follows that $1414\ \Omega = \sqrt{2\text{Re}^2[Z]} = \sqrt{2}\text{Re}[Z]$. Since $\sqrt{2} = 1.414$, we find that $\text{Re}[Z] = 1414\ \Omega/1.414 = \mathbf{10^3}\ \Omega$. The impedance is therefore given by

$$Z = 10^3\ \Omega - j10^3\ \Omega$$

This impedance is plotted in the following as a vector in the complex plane. Note the 'Australian' convention with the negative imaginary axis upwards.

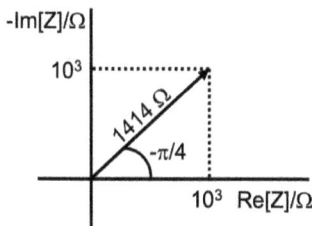

Problem 1.2

The admittance is given by $Y = \frac{1}{z} = \frac{1}{10^3\,\Omega - j10^3\,\Omega} = \frac{10^3\,\Omega + j10^3\,\Omega}{10^6\,\Omega^2 + 10^6\,\Omega^2} = \frac{10^{-3}}{2}\,\Omega^{-1} + j\frac{10^{-3}}{2}\,\Omega^{-1}$.

So that $Y = 5 \times 10^{-4}S + j5 \times 10^{-4}S$. The phase angle is now positive and equal to $\pi/4$. The magnitude of the admittance is given by (replacing Ω^{-1} by S).

$$|Y| = \sqrt{(5 \times 10^{-4})^2 + (5 \times 10^{-4})^2} = 7.071 \times 10^{-4}S$$

As expected, this is simply equal to $1/|z| = 1/1414\,\Omega$. The following plot shows the admittance vector in the complex plane – again Australian convention.

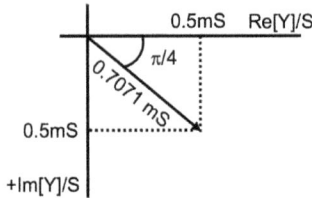

Problem 1.3

$$Z_C = -\frac{j}{\omega C} = -\frac{j}{2\pi f C} \qquad Z_L = +j\omega L = +2\pi f L$$

f/Hz	Z_c/j	Z_L/j
1	−1.592 MΩ	+6.282 mΩ
10	−159.2 kΩ	+62.82 mΩ
100	−15.92 kΩ	+628.2 mΩ
1000	−1.592 kΩ	+6.282 Ω

The following figure shows the Bode plots of magnitude vs. frequency.

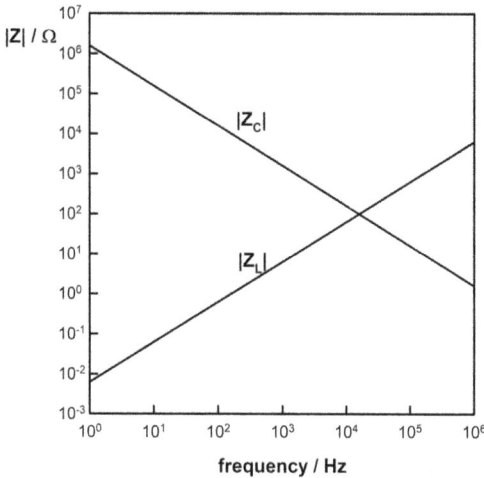

Problem 1.4

To find the crossing point where the two impedances have equal magnitude, we note that

$$2\pi FL = \frac{1}{2\pi fC} \quad f^2 = \frac{1}{4\pi^2 LC} \quad f = \frac{1}{2\pi\sqrt{LC}} = 15.92\,\text{kHz}$$

Problem 1.5

Looking at the circuit we can see that the total current will be composed of two contributions. One is the current through the second resistor, which is in parallel with the capacitor. This current will be independent of time and determined by the sum of the two resistances in the circuit, which is 2 kΩ. For a 1 V step, the current through the parallel resistor will be 0.5 mA. The second contribution to the current will be due to charging the parallel capacitor via the first resistor. This current will have an initial value that is determined by the first resistor, i.e., it will be 1 mA. The current through the capacitor will then fall exponentially towards zero with a time constant given by the RC product $10^3\Omega \times 10^{-6}\text{F} = 1$ ms. Adding the two current contributions gives the current transient shape as follows.

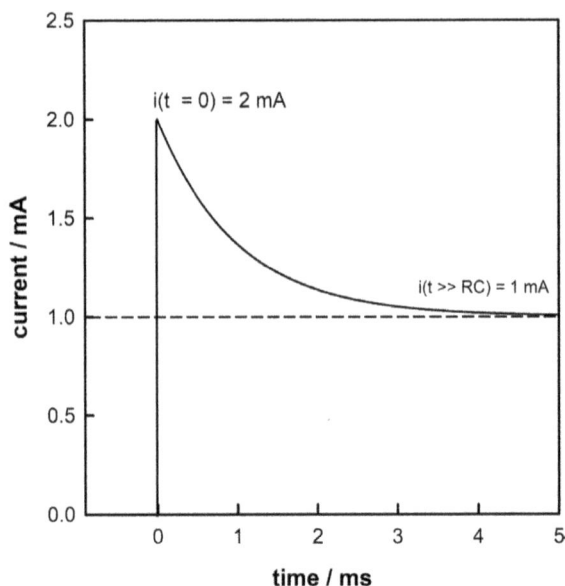

Problem 1.6

Again we can use common sense to work out what the current response will be for a series combination of resistor and inductor. At very short times, the inductor will block the current, so for the limit $t \to 0$, the current will be zero. In the limit $t \to \infty$, the current will be determined just by the resistor since the impedance of the inductor will be zero. So obviously, we have a current transient that rises exponentially to a maximum determined by the 100 Ω resistor, i.e., to 10 mA. What about the time constant of the rise? It helps to look at the units. Since $|Z_L| = \omega L$, the units of inductance (H) must be equivalent to Ωs. To obtain a time constant in seconds, we need to divide by Ω, so our time constant in this case is $\frac{L}{R} = \frac{10^{-3}\mathrm{H}}{100\,\Omega} = 10\,\mu$s. The current transient is therefore given by

$$i(t) = 10\,\mathrm{mA}(1 - e^{\frac{-t}{10^{-5}}})$$

The shape of the transient is as follows.

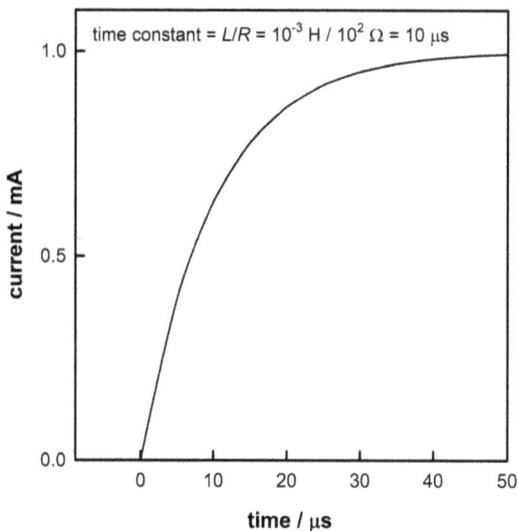

time constant = L/R = 10^{-3} H / 10^2 Ω = 10 μs

Chapter 3

Problem 3.1

The answer is quite easy. The definition of the hyperbolic cotangent is $\coth(x) = \frac{e^x + e^{-x}}{e^x - e^{-x}}$. In this problem, $x = \frac{l}{\lambda}$. If the thickness of the layer is much greater than the penetration depth ($l \gg \lambda$), x becomes large and you should be able to see that $\coth(x) \to 1$. The coth function is plotted in the following figure to confirm this. So the coth term disappears and we obtain the limit for the penetration depth being much less than the layer thickness.

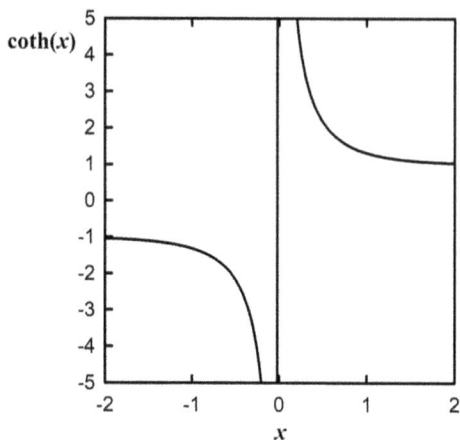

Index